CONTINUUM MECHANICS IN THE EARTH SCIENCES

Continuum mechanics lies at the heart of studies into many geological and geophysical phenomena, from earthquakes and faults to the fluid dynamics of the Earth. Based on the author's extensive teaching and research experience, this interdisciplinary book provides geoscientists, physicists and applied mathematicians with a class-tested, accessible overview of continuum mechanics.

Starting from thermodynamic principles and geometrical insights, the book surveys solid, fluid, and gas dynamics. In later review chapters, it explores new aspects of the field emerging from nonlinearity and dynamical complexity, and provides a brief introduction to computational modeling. Simple, yet rigorous, derivations are used to review the essential mathematics, assuming an undergraduate geophysics, physics, engineering or mathematics background. The author emphasizes the full three-dimensional geometries of real-world examples, enabling students to apply these in deconstructing solid-Earth and planet-related problems.

Problem sets and worked examples are provided, making this a practical resource for graduate students in geophysics, planetary physics and geology, as well as a beneficial tool for professional scientists seeking a better understanding of the mathematics and physics underlying problems in the Earth sciences.

WILLIAM I. NEWMAN is a Professor in the Department of Earth and Space Sciences as well as the Departments of Physics and Astronomy, and Mathematics at UCLA, where he has taught for over 30 years. A unifying feature of his research is the role of chaos and complexity in nature, in applications ranging from geophysics to astrophysics as well as mathematical, statistical, and computational modeling in condensed matter physics, climate dynamics, and theoretical biology. Professor Newman is a former member of the Institute for Advanced Study in Princeton, a John Simon Guggenheim Memorial Foundation Fellow, and has held appointments as the Stanislaw Ulam Distinguished Scholar at the Center for Nonlinear Studies, Los Alamos National Laboratory, and as the Morris Belkin Visiting Professor in Computational and Applied Mathematics at the Weizmann Institute of Science, Israel. For six months he also worked with colleagues in the Russian Academy of Sciences in Moscow on modeling earthquake events. He has served in an editorial capacity for the *Journal of Geophysical Research* and *Nonlinear Processes in Geophysics*, and has held elective office as chair of the Division of Dynamical Astronomy of the American Astronomical Society.

CONTINUUM MECHANICS IN THE EARTH SCIENCES

WILLIAM I. NEWMAN

CAMBRIDGE UNIVERSITY PRESS
Cambridge, New York, Melbourne, Madrid, Cape Town,
Singapore, São Paulo, Delhi, Mexico City

Cambridge University Press
The Edinburgh Building, Cambridge CB2 8RU, UK

Published in the United States of America by Cambridge University Press, New York

www.cambridge.org
Information on this title: www.cambridge.org/9780521562898

© William I. Newman 2012

This publication is in copyright. Subject to statutory exception
and to the provisions of relevant collective licensing agreements,
no reproduction of any part may take place without the written
permission of Cambridge University Press.

First published 2012

Printed in the United Kingdom at the University Press, Cambridge

A catalog record for this publication is available from the British Library

Library of Congress Cataloging in Publication data

ISBN 978-0-521-56289-8 Hardback

Cambridge University Press has no responsibility for the persistence or
accuracy of URLs for external or third-party internet websites referred to
in this publication, and does not guarantee that any content on such
websites is, or will remain, accurate or appropriate.

To my children, Ted and Jenny.

Contents

Preface		*page* ix
Acknowledgements		xii
1	**Some mathematical essentials**	1
	1.1 Scalars, vectors, and Cartesian tensors	1
	1.2 Matrices and determinants	7
	1.3 Transformations of Cartesian tensors	9
	1.4 Eigenvalues and eigenvectors	12
	1.5 Simplified approach to rotation	16
	1.6 Curvature, torsion, and kinematics	18
	Exercises	23
2	**Stress principles**	27
	2.1 Body and surface forces	27
	2.2 Cauchy stress principle	29
	2.3 Stress tensor	31
	2.4 Symmetry and transformation laws	33
	2.5 Principal stresses and directions	34
	2.6 Solving the cubic eigenvalue equation problem	37
	2.7 Maximum and minimum stress values	39
	2.8 Mohr's circles	42
	2.9 Plane, deviator, spherical, and octahedral stress	44
	Exercises	46
3	**Deformation and motion**	49
	3.1 Coordinates and deformation	49
	3.2 Strain tensor	53
	3.3 Linearized deformation theory	54
	3.4 Stretch ratios	58
	3.5 Velocity gradient	59
	3.6 Vorticity and material derivative	61
	Exercises	64

4	**Fundamental laws and equations**	67
	4.1 Terminology and material derivatives	67
	4.2 Conservation of mass and the continuity equation	71
	4.3 Linear momentum and the equations of motion	73
	4.4 Piola–Kirchhoff stress tensor	74
	4.5 Angular momentum principle	75
	4.6 Conservation of energy and the energy equation	76
	4.7 Constitutive equations	79
	4.8 Thermodynamic considerations	82
	Exercises	86
5	**Linear elastic solids**	89
	5.1 Elasticity, Hooke's law, and free energy	89
	5.2 Homogeneous deformations	94
	5.3 Role of temperature	97
	5.4 Elastic waves for isotropic bodies	100
	5.5 Helmholtz's decomposition theorem	102
	5.6 Statics for isotropic bodies	104
	5.7 Microscopic structure and dislocations	106
	Exercises	109
6	**Classical fluids**	112
	6.1 Stokesian and Newtonian fluids: Navier–Stokes equations	113
	6.2 Some special fluids and flows	117
	Exercises	129
7	**Geophysical fluid dynamics**	134
	7.1 Dimensional analysis and dimensionless form	134
	7.2 Dimensionless numbers	138
	Exercises	143
8	**Computation in continuum mechanics**	147
	8.1 Review of partial differential equations	148
	8.2 Survey of numerical methods	153
9	**Nonlinearity in the Earth**	159
	9.1 Friction	163
	9.2 Fracture	165
	9.3 Percolation and self-organized criticality	168
	9.4 Fractals	171
	References	175
	Index	180

Preface

This book is the outcome of an introductory graduate-level course that I have given at the University of California, Los Angeles for a number of years as part of our program in Geophysics and Space Physics. Our program is physics-oriented and draws many of its students from the ranks of undergraduate physics and, sometimes, mathematics majors, in addition to geophysics and occasionally geology majors. Accordingly, this text approaches the subject by promoting a physics-based understanding of the basic principles with a relatively rigorous mathematical approach. Since the needs of this course were rather unique, blending concepts in physics and mathematics with the Earth sciences, I approached teaching the subject by drawing on many sources in developing the necessary material. (Throughout this volume, I refer to materials that provide more complete treatments of the topics which we only have time to overview.) In contrast to other sources, I wanted this course to treat not only classical methods but survey some of the ideas emerging in the geosciences that were drawn directly from current ideas in physics, especially nonlinear dynamics. Over time, the material developed more coherence and my lecture notes for this academic quarter-long course evolved into this text.

The subject of continuum mechanics is predicated on the notion that many natural phenomena have a fundamentally smooth, continuous nature. This constitutes the basis for solid and fluid mechanics, major components of this course. However, there are also some important departures from the continuum assumption which must be reviewed and this is done in the later chapters. Thus, this text surveys a mixture of traditional as well as contemporary topics.

I begin by introducing the mathematical notation, stress principles, and kinematic description that is common to almost all books on this subject. However, I emphasize the three-dimensionality that prevails in many geoscience problems and the techniques needed to accommodate that geometric complexity. The fundamental laws and equations are introduced employing a thermodynamic basis for their derivation, rather than simply assuming an underlying linearity, as is done in most engineering texts. Mathematical derivations are provided where they have proven to be helpful during lecture presentations. Linear elasticity is then introduced, covering many issues encountered in statics and dynamics, including the equations for seismic wave propagation. Exercises are provided at the end of each of these chapters.

Since this graduate course was designed to be given during an academic quarter, rather than a semester, I chose to develop four additional chapters to survey a number of themes that emerge from continuum mechanics that could be introduced if sufficient time were available. (During a semester-long offering, the four added survey chapters can be augmented by readings from the various books and other materials cited.) Because each of these chapters present brief surveys, I preface each of them by citing representative books in the literature that can provide much greater detail. Classical fluid flow is introduced, including some elements of perturbation theory as well as the application of eddy viscosities and the scaling laws often encountered in turbulent flows. Additional attention is placed on geophysical fluid flows and their associated power laws, and we consider how complications in the underlying physics can sometimes break these simple scalings. Since the complexity of these real-world problems often compels us to use the computer to simulate these continua, the penultimate chapter provides a brief survey of some of the current computational methodologies available, so that students pursuing the current literature will have a basic understanding of how numerical results are obtained. Finally, I go on to survey a number of themes germane to nonlinearity in the Earth, beginning with some of the phenomenology of earthquakes, including self-similar scaling and fractals, and some current thinking on models including percolation and self-organized criticality. While not designed to be an in-depth introduction, these additional chapters have been constructed to provide graduate students relatively early in their studies a taste for some of the topics that have grown to be important. At the same time, this material is designed to help them learn enough of the prevailing ideas to be able to benefit from seminars and other venues where these topics are discussed. While the final two chapters do not have exercises provided, they have proven useful in relating some of the more conventional topics associated with continuum mechanics to real geoscience problems and, therefore, help provide an entrée for graduate students into these arenas.

The Earth and space environment provide both an important testing ground for many ideas in continuum mechanics, as well as a cogent need for the development of this subject. Indeed, it is the art of bridging the physics and mathematics of continuum mechanics with their application to the geosciences that poses the greatest challenge and provides the greatest benefits.

While completing this volume, my editors at Cambridge University Press, Susan Francis and Laura Clark, asked me to provide a cover image for this book and recommended that a photograph be adopted, instead of a geometrical design or blank cover as is so often employed in technical books. For this purpose, I selected a photograph that my daughter, Jenny, and I had taken several years ago of Devils Tower in northeastern Wyoming. This picture provides a beautiful illustration of the competition of two very different processes encountered in continuum mechanics in the formation of this remarkable geologic feature. Devils Tower is an igneous intrusion that rises 386 m above the surrounding sedimentary terrain and the Belle Fourche River. As the basalt cooled, the outer surface cooled at a more rapid pace than the flow material inside, and thermal conduction from the interior could not keep up with the exterior cooling and circumferential contraction. Columnar fractures and joints emerged during the cooling process and yielded a set of parallel prismatic columns presenting a hexagonal pattern. A number of aspects relating to the formation of Devils Tower remain controversial. Nevertheless, this American national monument serves as a striking example of the interplay between the different kinds of geology, physics, mathematics, and geometry that define continuum mechanics.

Acknowledgements

I gratefully acknowledge the advice and comments provided by the many students to whom I have had the privilege of teaching this material. Their comments and advice were invaluable. I have benefited as well from the advice of many of my colleagues, notably Donald Turcotte, Andrei Gabrielov, Jonathan Aurnou, Gerald Schubert, Vladimir Keilis-Borok, Gerhard Oertel, Paul Davis, and Peter Bird, among others. I also wish to thank William Kaula and Leon Knopoff, who had a profound influence upon my appreciation for this subject. I want to thank several of our graduate students who took time out to help proofread the manuscript, especially Nathaniel Hamlin, Jon Harrington, Shantanu Naidu, Sebastiano Padovan, and Igor Stubailo. Finally, I wish to thank my editors at Cambridge University Press, Susan Francis and Laura Clark, for their steadfast support.

1
Some mathematical essentials

1.1 Scalars, vectors, and Cartesian tensors

Geometry is a vital ingredient in the description of continuum problems. Our treatment will focus on the mathematically simplest representation for this subject. Although curvilinear coordinates can be more natural, they introduce complications that go beyond the scope of this book. The initial part of our treatment will parallel the Cartesian approach of Mase and Mase (1990) rather than the curvilinear approach of Narasimhan (1993) and Fung (1965). Hence, we will adhere to a Cartesian description of problems and be spared the need to distinguish between covariant and contravariant notation. Moreover, we will generally employ second-rank tensors which are matrices that possess some very special and important (coordinate) transformation properties.

We will distinguish between three classes of objects: namely, scalars, vectors, and tensors. In reality, all quantities may be regarded as tensors of a specific rank. *Scalar* (nonconstant) quantities, such as density and temperature, are *zero rank* or *order* tensors, while *vector* quantities (which have an associated direction, such as velocity) are *first-rank* tensors. *Second-rank tensors*, such as the stress tensor, are a special case of square matrices. We will usually denote vector quantities by bold-face lower-case letters, while second-rank tensors will be denoted by bold-face upper-case letters.

To simplify our geometrical description of problems, we will employ an indicial notation. In lieu of x, y, and z orthogonal axes, we will employ x_1, x_2, and x_3. Similarly, we will denote by $\hat{\mathbf{e}}_1$, $\hat{\mathbf{e}}_2$, and $\hat{\mathbf{e}}_3$ *unit*-vectors in the directions of x_1, x_2, and x_3. The indicial notation implies that any repeated index is implicitly summed, generally from 1 through 3. This is the *Einstein summation convention*. It is sufficient to denote a vector $\mathbf{v}$ by its three components (v_1, v_2, v_3). We conform with the tradition that vector quantities are shown in bold face. For more detail on these issues, the reader is advised to consult mathematical texts such as those by Boas

(2006), Arfken and Weber (2005), Mathews and Walker (1970), or Schutz (1980) ranging from elementary to advanced treatments. We note that **v** can be represented vectorially then as

$$\mathbf{v} \equiv \sum_{i=1}^{3} v_i \,\hat{\mathbf{e}}_i = v_i \,\hat{\mathbf{e}}_i. \tag{1.1}$$

Accordingly, we see that

$$T_{ij}\, v_i \,\hat{\mathbf{e}}_j = \sum_{i=1, j=1}^{3} T_{ij} v_i \hat{\mathbf{e}}_j. \tag{1.2}$$

Moreover, vectors and matrices have certain of the properties of a group: (a) the sum of two vectors is a vector; (b) the product of a vector by a scalar is a vector, etc. In addition, we define an inner (scalar) product or dot product according to the usual physicist's convention

$$\mathbf{u} \cdot \mathbf{v} \equiv u_i\, v_i. \tag{1.3}$$

It is important to note that mathematicians sometimes employ a slightly different notation for the inner product, namely

$$(\mathbf{u}, \mathbf{v}) = \mathbf{u}^T \mathbf{v} = \mathbf{v}^T \mathbf{u} = u_i\, v_i, \tag{1.4}$$

where they assume that **u** and **v** are column vectors. The latter formalism requires particular care since transposed quantities often appear. In order to maintain transparency in all of our derivations, we will employ primarily the indicial notation. Moreover, if we define u and v to be the lengths of **u** and **v**, respectively, according to

$$u \equiv \sqrt{u_i\, u_i} = |\mathbf{u}|; \qquad v \equiv \sqrt{v_i\, v_i} = |\mathbf{v}|, \tag{1.5}$$

we can identify an angle θ between **u** and **v** which we define according to

$$\mathbf{u} \cdot \mathbf{v} \equiv u\, v \cos \theta. \tag{1.6}$$

We now introduce the Kronecker delta δ_{ij} and the Levi-Civita permutation symbol ϵ_{ijk} owing to their utility in tensor calculations.

We define the Kronecker delta according to

$$\delta_{ij} \equiv \begin{cases} 1, & \text{if } i = j \\ 0, & \text{if } i \neq j \end{cases}. \tag{1.7}$$

It follows that the Kronecker delta is the realization of the identity matrix. It follows then that

$$\hat{\mathbf{e}}_i \cdot \hat{\mathbf{e}}_j = \delta_{ij}, \tag{1.8}$$

and that
$$\delta_{ii} = 3. \tag{1.9}$$
(This is equivalent to saying that the *trace*, i.e. the sum of the diagonal elements, of the identity matrix is 3.) An important consequence of equation (1.8) is that
$$\delta_{ij}\,\hat{\mathbf{e}}_j = \hat{\mathbf{e}}_i. \tag{1.10}$$
We can employ these definitions to derive the general scalar product relation (1.3) using the special case (1.8). In particular, it follows that
$$\mathbf{u}\cdot\mathbf{v} = u_i\,\hat{\mathbf{e}}_i \cdot v_j\,\hat{\mathbf{e}}_j = u_i\,v_j\,\hat{\mathbf{e}}_i\cdot\hat{\mathbf{e}}_j = u_i\,v_j\,\delta_{ij} = u_i\,v_i. \tag{1.11}$$

In order to introduce the vector (cross) product, we introduce the Levi-Civita or permutation symbol ϵ_{ijk} according to
$$\epsilon_{ijk} \equiv \begin{cases} 1, & \text{if } i\,j\,k \text{ are an even permutation of } 1\,2\,3 \\ -1, & \text{if } i\,j\,k \text{ are an odd permutation of } 1\,2\,3 \\ 0, & \text{if any two of } i,\,j,\,k \text{ are the same} \end{cases}. \tag{1.12}$$

We note that ϵ_{ijk} changes sign if any two of its indices are interchanged. For example, if the 1 and 3 are interchanged, then the sequence 1 2 3 becomes 3 2 1. Accordingly, we *define* the cross product $\mathbf{u}\times\mathbf{v}$ according to its ith component, namely
$$(\mathbf{u}\times\mathbf{v})_i \equiv \epsilon_{ijk}\,u_j\,v_k, \tag{1.13}$$
or, equivalently,
$$\mathbf{u}\times\mathbf{v} = (\mathbf{u}\times\mathbf{v})_i\,\hat{\mathbf{e}}_i = \epsilon_{ijk}\,\hat{\mathbf{e}}_i\,u_j\,v_k = -(\mathbf{v}\times\mathbf{u}). \tag{1.14}$$

By inspection, it is observed that this kind of structure is closely connected to the definition of the determinant of a 3×3 matrix. This follows directly when we write the scalar triple product
$$\mathbf{u}\cdot(\mathbf{v}\times\mathbf{w}) = \epsilon_{ijk}\,u_i\,v_j\,w_k, \tag{1.15}$$
which, by virtue of the cyclic permutivity of the Levi-Civita symbol demonstrates
$$\mathbf{u}\cdot(\mathbf{v}\times\mathbf{w}) = \mathbf{v}\cdot(\mathbf{w}\times\mathbf{u}) = \mathbf{w}\cdot(\mathbf{u}\times\mathbf{v}). \tag{1.16}$$

The right side of equation (1.15) is the determinant of a matrix whose rows correspond to $\mathbf{u}$, $\mathbf{v}$, and $\mathbf{w}$, respectively. It is useful to note that the scalar triple product can be employed to establish whether a Cartesian coordinate system is right or left handed, i.e. the product is $+1$ or -1.

It is useful, as well, to consider the vector triple cross product
$$\mathbf{u}\times(\mathbf{v}\times\mathbf{w}) = \mathbf{u}\times\epsilon_{ijk}\,\hat{\mathbf{e}}_i\,v_j\,w_k = \epsilon_{lmi}\,\hat{\mathbf{e}}_l\,u_m\,\epsilon_{ijk}\,v_j\,w_k$$
$$= \left(\epsilon_{ilm}\,\epsilon_{ijk}\right)\hat{\mathbf{e}}_l\,u_m\,v_j\,w_k. \tag{1.17}$$

It is necessary to deal first with the $\epsilon_{ilm}\epsilon_{ijk}$ term. Observe, as we sum over the i index, that contributions can emerge only if $l \neq m$ and $j \neq k$. If these conditions both hold, then we get a contribution of 1 if $l = j$ and $m = k$ and a contribution of -1 if $l = k$ and $m = j$. Hence, it follows that

$$\epsilon_{ilm}\,\epsilon_{ijk} = \delta_{lj}\,\delta_{mk} - \delta_{lk}\,\delta_{mj}. \tag{1.18}$$

Returning to equation (1.17), we obtain that

$$\begin{aligned}
\mathbf{u} \times (\mathbf{v} \times \mathbf{w}) &= \left(\delta_{lj}\,\delta_{mk} - \delta_{lk}\,\delta_{mj}\right) \hat{\mathbf{e}}_l\, u_m\, v_j\, w_k \\
&= \hat{\mathbf{e}}_l\, v_l\, u_m\, w_m - \hat{\mathbf{e}}_l\, w_l\, u_m\, v_m \\
&= \mathbf{v}\,(\mathbf{u} \cdot \mathbf{w}) - \mathbf{w}\,(\mathbf{u} \cdot \mathbf{v}),
\end{aligned} \tag{1.19}$$

thereby reproducing a familiar, albeit otherwise cumbersome to derive, algebraic identity. Finally, if we replace the role of $\mathbf{u}$ in the triple scalar product (1.15) by $\mathbf{v} \times \mathbf{w}$, it immediately follows that

$$\begin{aligned}
(\mathbf{v} \times \mathbf{w}) \cdot (\mathbf{v} \times \mathbf{w}) &= |\mathbf{v} \times \mathbf{w}|^2 = \epsilon_{ijk}\, v_j\, w_k \epsilon_{ilm}\, v_l\, w_m \\
&= \left(\delta_{jl}\delta_{km} - \delta_{jm}\delta_{kl}\right) v_j\, w_k\, v_l\, w_m.
\end{aligned} \tag{1.20}$$

Finally, this can be written

$$|\mathbf{v} \times \mathbf{w}|^2 = v^2 w^2 - (\mathbf{v} \cdot \mathbf{w})^2 = v^2\, w^2\, \sin^2\theta, \tag{1.21}$$

where we have made use of the definition (1.6). Indeed, it is possible to demonstrate the validity of many other vector identities by employing the Levi-Civita and Kronecker symbols. This is especially true with respect to *derivative* operators. We define ∂_i according to

$$\partial_i \equiv \frac{\partial}{\partial x_i}. \tag{1.22}$$

Another notational shortcut that is commonly used is to employ a subscript of ", i" to denote a derivative with respect to x_i; importantly, a comma "," is employed to designate differentiation together with the subscript. Hence, if f is a scalar function of $\mathbf{x}$, we write

$$\frac{\partial f}{\partial x_i} = \partial_i f = f_{,i}; \tag{1.23}$$

but if $\mathbf{g}$ is a vector function of $\mathbf{x}$, we write

$$\frac{\partial g_i}{\partial x_j} = \partial_j\, g_i = g_{i,j}. \tag{1.24}$$

Higher derivatives may be expressed using this shorthand as well, e.g.

$$\frac{\partial^2 g_i}{\partial x_j\, \partial x_k} = g_{i,jk}. \tag{1.25}$$

1.1 Scalars, vectors, and Cartesian tensors

Then, the usual gradient, divergence, and curl operators become

$$\nabla = \hat{\mathbf{e}}_i \, \partial_i; \tag{1.26}$$

$$\nabla \cdot \mathbf{u} = \partial_i \, u_i; \tag{1.27}$$

and

$$\nabla \times \mathbf{u} = \epsilon_{ijk} \, \hat{\mathbf{e}}_i \, \partial_j \, u_k, \tag{1.28}$$

where $\mathbf{u}$ is a vector and a function of the position vector $\mathbf{x}$. With only a modest degree of additional effort (noting that ∂_i commutes with ∂_j, an added benefit of Cartesian geometry, but not with any *function* of $\mathbf{x}$), it is now relatively simple to derive all vector identities in Cartesian coordinates. Before proceeding further, it deserves mention that the usual theorems of Gauss, Green, and Stokes also hold for tensor quantities albeit in a slightly more complicated form than for vector ones. The essential point here is that as one converts from a volume integral to a surface integral and to a line integral, appropriate differential operators are introduced. In a Cartesian representation, these rarely cause any problems and the rules discussed above apply.

One final notation issue needs to be addressed at this time, the tensor (outer) product of two vectors, the so-called *dyad*

$$\mathbf{u}\,\mathbf{v} = u_i \, \hat{\mathbf{e}}_i \, v_j \, \hat{\mathbf{e}}_j = u_i \, v_j \, \hat{\mathbf{e}}_i \, \hat{\mathbf{e}}_j. \tag{1.29}$$

(Although some engineering texts sometimes introduce the $\otimes$ symbol between the two vector quantities, mathematics texts often employ the $\otimes$ symbol to denote an "antisymmetric" form, i.e. $\mathbf{u} \otimes \mathbf{v} = \mathbf{u}\mathbf{v} - \mathbf{v}\mathbf{u}$. Other mathematics texts employ the "wedge" $\mathbf{u} \wedge \mathbf{v}$ for this purpose. The absence of an intervening dot between the unit vectors is significant: there is no inner product implied.) Operations on dyads follow the usual rules for the relevant vector components; e.g.

$$\mathbf{u}\,\mathbf{v} \cdot \mathbf{p}\,\mathbf{q} = \mathbf{u}\,(\mathbf{v} \cdot \mathbf{p})\,\mathbf{q} = (\mathbf{v} \cdot \mathbf{p})\,\mathbf{u}\,\mathbf{q}, \tag{1.30}$$

where we note that the scalar product $(\mathbf{v} \cdot \mathbf{p})$ should be regarded solely as a scalar quantity. Dyads are particularly useful in the decomposition or diagonalization of matrices. Note, also, that although a dyad has nine components, only six independent quantities are involved (and these six quantities can be calculated up to a multiplicative constant). It is possible in similar fashion to construct triadic, tetradic, and higher rank tensors.

Thus far, our treatment of geometry and some of the underpinnings of scalars, vectors and Cartesian tensors has been abstract. The power of these methods is greatly enhanced when we employ our geometric intuition in solving problems. (This philosophy is also at the heart of our adherence to Cartesian coordinates in our treatment. Once we have derived the fundamental equations of continuum mechanics in Cartesian form, it is relatively simple to convert them to other curvilinear coordinate systems, such as cylindrical or spherical. What we gain in the process is geometrical simplicity.)

As an illustration of the utility of coupling geometric intuition with the formalism, consider the classic problem of establishing the bond angles in a methane or CH_4 molecule. Visualize the carbon atom at the origin of our coordinate system and let us assume that one of the hydrogen atoms is at a distance v from the carbon atom along the z-axis. We will designate its position by the vector $\mathbf{v}^{(0)}$. The three remaining hydrogen atoms make an angle θ, to be determined, with respect to the hydrogen atom on the z-axis, and we will call their vectors $\mathbf{v}^{(i)}$ for $i = 1, 2, 3$. The "center of mass" of the hydrogen atoms is $\sum_{i=0}^{3} \mathbf{v}^{(i)} = \mathbf{0}$, i.e. at the origin where the carbon atom is situated. We take the dot product of $\mathbf{v}^{(0)}$ with this latter quantity to find that $v^2 + 3 v^2 \cos \theta = 0$ leaving $\cos \theta = -1/3$ or $\theta \approx 109.471\,220\,634°$. Drawing a picture is *always* a good idea. Let us consider a more complex example.

As a novel illustration of the use of dyads, consider the "corner cube" reflector frequently employed on highways and elsewhere owing to their remarkable ability to reflect back to the source light incident on the device from *any* angle.

A corner reflector consists of three mutually perpendicular, intersecting flat surfaces (see figure 1.1). In particular, assume that the corner's faces can be described by normal unit vectors $\hat{\mathbf{n}}_i$, $i = 1, 2, 3$ as shown in the accompanying figure. If a light ray with direction $\hat{\mathbf{r}}$ impinges on face #1, then its projection along the $\hat{\mathbf{n}}_1$ direction is reversed while its projection on the $\hat{\mathbf{n}}_i$ for $i = 2, 3$ remains unchanged. Thus, the new vector $\hat{\mathbf{r}}'$ after the first reflection is given by

$$\hat{\mathbf{r}}' = \left(\hat{\mathbf{n}}_2 \, \hat{\mathbf{n}}_2 + \hat{\mathbf{n}}_3 \, \hat{\mathbf{n}}_3 - \hat{\mathbf{n}}_1 \, \hat{\mathbf{n}}_1 \right) \cdot \hat{\mathbf{r}} = \left(\mathbf{I} - 2 \hat{\mathbf{n}}_1 \, \hat{\mathbf{n}}_1 \right) \cdot \hat{\mathbf{r}}, \tag{1.31}$$

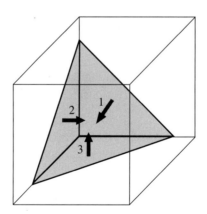

Figure 1.1 Geometry of a corner cube retroreflector.

where

$$\mathbf{I} = \hat{\mathbf{n}}_1\,\hat{\mathbf{n}}_1 + \hat{\mathbf{n}}_2\,\hat{\mathbf{n}}_2 + \hat{\mathbf{n}}_3\,\hat{\mathbf{n}}_3 \tag{1.32}$$

defines the identity operator. We now allow for a reflection on face number 2, yielding $\hat{\mathbf{r}}''$. Taking all three reflections in turn, we obtain for the final ray vector

$$\hat{\mathbf{r}}''' = \left(\mathbf{I} - 2\,\hat{\mathbf{n}}_3\,\hat{\mathbf{n}}_3\right) \cdot \left(\mathbf{I} - 2\,\hat{\mathbf{n}}_2\,\hat{\mathbf{n}}_2\right) \cdot \left(\mathbf{I} - 2\,\hat{\mathbf{n}}_1\,\hat{\mathbf{n}}_1\right) \cdot \hat{\mathbf{r}} = -\hat{\mathbf{r}}, \tag{1.33}$$

after a modest amount of algebra where the scalar products of the form $\hat{\mathbf{n}}_1 \cdot \hat{\mathbf{n}}_2$, etc., are observed to vanish. Hence, we see how the utilization of these geometrical constructs can dramatically simplify the solution of practical problems.[1]

The notation employed thus far has mixed symbolic and indicial conventions; the proper treatment requires a covariant–contravariant formulation which would also accommodate curvilinear coordinates. Engineering texts generally employ notation that preserves vectorial quantities using bold-face characters, e.g. $\mathbf{n}$, while physics texts such as Landau *et al.* (1986) and Landau and Lifshitz (1987) generally employ indicial notation, e.g. n_i. When referring to tensors (and to matrices), the symmetries possessed with respect to the indices can be particularly important. For example, if $A_{ij} = -A_{ji}$, we say that the second-rank tensor $\mathbf{A}$ is anti-symmetric; similarly, if $A_{ij} = A_{ji}$, we say that $\mathbf{A}$ is symmetric. To maintain a (subtle) distinction between matrices and tensors, we shall denote the associated matrix by the symbol $\mathcal{A}$. We turn now to some of the essential properties of matrices and determinants.

1.2 Matrices and determinants

A *matrix*, unlike a tensor which properly defined also preserves coordinate transformation properties, is an ordered rectangular array of elements enclosed by brackets. The reader may wish to review a good linear algebra text before continuing further. An encyclopedic source of information concerning matrices is the two-volume text by Gantmakher (1959). An M by N matrix (written $M \times N$) can be expressed

$$\mathcal{A} = \left[A_{ij}\right] = \begin{pmatrix} A_{11} & A_{12} & \ldots & A_{1N} \\ A_{21} & A_{22} & \ldots & A_{2N} \\ \vdots & \vdots & \ddots & \vdots \\ A_{M1} & A_{M2} & \ldots & A_{MN} \end{pmatrix}. \tag{1.34}$$

---

[1] An especially elegant way of demonstrating the reversal of the direction of the incident beam is to adopt temporarily a rotated coordinate system whose axes are orthogonal to each of the three cube faces. In this system, each of the components of the ray vector undergoes a reversal as the light ray encounters each cube face, respectively. Since all three components of the initial vector are reversed after the three reflections, the outcome of this encounter with a corner cube is the reversal of the light ray.

A *zero* or *null* matrix has all elements equal to zero. A *diagonal* matrix is a square matrix whose elements which are not on the *principal diagonal* vanish; the *unit* or *identity* matrix is a diagonal matrix whose diagonal elements are unit values. By interchanging the rows and columns of an $M \times N$ matrix $\mathcal{A}$, we form its $N \times M$ transpose $\mathcal{A}^T$. It follows then that a symmetric matrix is a square matrix $\mathcal{A} = \mathcal{A}^T$. Any matrix can be expressed as the sum of a symmetric and an antisymmetric matrix, respectively, namely

$$\mathcal{A} = \frac{\mathcal{A} + \mathcal{A}^T}{2} + \frac{\mathcal{A} - \mathcal{A}^T}{2}. \tag{1.35}$$

For complex-valued matrices, we sometimes denote by the superscript † or H the complex conjugate transpose or *Hermitian* operator. Thus, we define $\mathcal{A}^\dagger \equiv \mathcal{A}^H \equiv \mathcal{A}^{T\star}$ where a $\star$ is used to indicate the complex conjugate. We say that a matrix $\mathcal{A}$ is Hermitian if it is identical to its conjugate transpose. Such properties sometimes emerge in the manipulation of matrices, but rarely so in continuum mechanics.

Matrix addition is commutative, i.e. $\mathcal{A} + \mathcal{B} = \mathcal{B} + \mathcal{A}$, and associative, i.e. $\mathcal{A} + (\mathcal{B} + \mathcal{C}) = (\mathcal{A} + \mathcal{B}) + \mathcal{C}$. Multiplication of a matrix $\mathcal{A}$ by a scalar λ gives rise to a new matrix $\lambda \mathcal{A}$. A matrix product $\mathcal{C} = \mathcal{A}\mathcal{B}$ can be defined in a manner reminiscent of an inner product, that is

$$C_{ij} = A_{ik} B_{kj}, \tag{1.36}$$

where summation over the index k takes place over the admissible range – note that the matrices $\mathcal{A}$ and $\mathcal{B}$ must be compatible in size. Observe further that matrix multiplication is *not* commutative, i.e. $\mathcal{A}\mathcal{B} \neq \mathcal{B}\mathcal{A}$.

It is often useful to define a *quadratic form* from a matrix $\mathcal{A}$ or its equivalent second rank tensor **A**, namely $\mathbf{x}^H \mathcal{A} \mathbf{x}$. We say that a matrix is positive definite if $\mathbf{x}^H \mathcal{A} \mathbf{x} > 0$ for all $\mathbf{x}$ and that it is positive semi-definite if $\mathbf{x}^H \mathcal{A} \mathbf{x} \geq 0$. (This property of matrices is of fundamental importance in continuum mechanics and in stability theory.) The sum of the diagonal elements of a matrix is its trace, i.e. $\text{tr } \mathcal{A} \equiv A_{ii}$. For convenience, we will define the determinant of a matrix using the Levi-Civita symbol, extending the definition (1.12) of the symbol to an arbitrary number N of indices (corresponding to an $N \times N$ matrix). In particular, we define

$$\epsilon_{i_1 i_2 \ldots i_N} \equiv \begin{cases} 1, & \text{if } i_1 i_2 \ldots i_N \text{ are an even permutation of } 1, 2, \ldots N \\ -1, & \text{if } i_1 i_2 \ldots i_N \text{ are an odd permutation of } 1, 2, \ldots N. \\ 0, & \text{if any two of } i_1 i_2 \ldots i_N \text{ are the same} \end{cases} \tag{1.37}$$

The issue of even and odd permutations can be understood this way: if we interchange any two indices of the Levi-Civita symbol, then it changes sign, alternating between $+1$ and -1. Accordingly, however, we note that if any two indices are the same, then the Levi-Civita symbol must vanish. Then, we define $\det \mathcal{A}$ according to

$$\det \mathcal{A} \equiv \epsilon_{i_1 i_2 \ldots i_N} A_{1 i_1} A_{2 i_2} \ldots A_{N i_N}. \tag{1.38}$$

It is easy to show that this reduces to the familiar definition for the determinant of a 3×3 matrix. Moreover, it is easy to show that all of the familiar properties of a determinant are preserved through this definition – particularly, if we regard each row (or column) of $\mathcal{A}$ as a vector, then the determinant vanishes if any row (or column) can be expressed as a linear combination of the other rows (or columns). Finally, these rules can be employed to develop the usual rules for the evaluation of a determinant by cofactors, etc. The inverse of a matrix $\mathcal{A}^{-1}$, *if it exists*, is a matrix defined such that $\mathcal{A}^{-1} \mathcal{A} = \mathcal{A} \mathcal{A}^{-1} = \mathcal{I}$, the identity matrix. We observe that the transpose or inverse of a product reverses the usual order of terms, that is

$$(\mathcal{A}\mathcal{B})^T = \mathcal{B}^T \mathcal{A}^T, \tag{1.39}$$

and

$$(\mathcal{A}\mathcal{B})^{-1} = \mathcal{B}^{-1} \mathcal{A}^{-1}. \tag{1.40}$$

Also, it is important to note that the determinant of a product is equal to the product of the determinants, namely

$$\det (\mathcal{A}\mathcal{B}) = (\det \mathcal{A}) (\det \mathcal{B}). \tag{1.41}$$

Although harder to prove, this is demonstrable using the Levi-Civita expression for the determinant of a matrix.

1.3 Transformations of Cartesian tensors

It is possible to convert (or rotate from) one coordinate system to another via a "transformation" – the existence of this transformation for tensors but not in general for matrices is what makes tensors a special case of matrices. Suppose we have one set of coordinate axes defined by unit vectors $\hat{\mathbf{e}}_i$ and wish to transform to another set of coordinate axes defined by unit vectors $\hat{\mathbf{e}}'_i$. Then, we can write (since we are dealing with a linear superposition)

$$\hat{\mathbf{e}}'_i = A_{ij} \hat{\mathbf{e}}_j, \tag{1.42}$$

where it follows that

$$A_{ij} = \hat{\mathbf{e}}'_i \cdot \hat{\mathbf{e}}_j. \tag{1.43}$$

Thus, we see that the coefficients of the transformation matrix are just the "direction cosines" defining the angles between the old and new coordinate systems. Expression (1.42) will be employed universally in transforming vectors and, later, tensors from one Cartesian coordinate system to another. Similarly, we can write for the inverse transform

$$\hat{\mathbf{e}}_i = A_{ij}^{-1} \hat{\mathbf{e}}'_j, \tag{1.44}$$

where

$$A_{ij}^{-1} = \hat{\mathbf{e}}_i \cdot \hat{\mathbf{e}}'_j. \tag{1.45}$$

By inspection, we observe that

$$\mathcal{A}^{-1} = \mathcal{A}^T. \tag{1.46}$$

The action of $\mathcal{A}$ on a vector can be regarded, alternately, as a rotation from one coordinate system to another (i.e. from the unprimed to the primed), or as a physical rotation of the vector in a manner specified by the reorientation of the coordinate axes.

The matrix $\mathcal{A}$ has some notable properties and, in particular, is referred to as *unitary* or length preserving. This can be shown by observing that, if

$$\mathbf{u}' \equiv \mathbf{A} \cdot \mathbf{u} = \mathcal{A}\mathbf{u}, \tag{1.47}$$

or, using indicial notation,

$$u'_i = A_{ij} u_j, \tag{1.48}$$

then

$$u'^2 = {\mathbf{u}'}^T \cdot \mathbf{u}' = \mathbf{u}^T \cdot \mathbf{A}^T \cdot \mathbf{A} \cdot \mathbf{u} = \mathbf{u}^T \cdot \mathbf{u} = u^2, \tag{1.49}$$

for *any* $\mathbf{u}$. Another important outcome of equations (1.42) and (1.48) emerges due to

$$\mathbf{u}' = u'_i \hat{\mathbf{e}}'_i = A_{ij} u_j A_{ik} \hat{\mathbf{e}}_k = A^T_{ji} A_{ik} u_j \hat{\mathbf{e}}_k = \delta_{jk} u_j \hat{\mathbf{e}}_k = \mathbf{u}, \tag{1.50}$$

thereby showing that the two representations that we have for our original vector $\mathbf{u}$ are identical. The transformation equation (1.48) allows us to convert almost effortlessly from one coordinate system to the other without changing any physical quantities.

If we regard each column of $\mathcal{A}$ as being a vector, say $\mathbf{u}_{(i)}$, $i = 1, 2, 3$, a little manipulation produces (since the inverse and transpose of $\mathcal{A}$ are identical) the result

$$\mathbf{u}_{(i)} \cdot \mathbf{u}_{(j)} = \delta_{ij}. \tag{1.51}$$

Thus, these three vectors are both orthogonal and of unit length. Similarly, one can regard each row of $\mathcal{A}$ as being a vector, say $\mathbf{v}_{(i)}$, and similarly show that these three vectors are also orthogonal and of unit length. Note that we used inner-product notation in the context of a tensor, i.e., we considered the vector $\mathbf{A} \cdot \mathbf{u}$ which is functionally equivalent to calculating $A_{ij} u_j$. If we had another tensor, say **B**, we employ a colon ":" to describe summation over two indices, i.e. $\mathbf{A} : \mathbf{B} = A_{ij} B_{ji}$. We will utilize this notation in the fourth chapter.

It is easy to prove that the product of two rotations is also unitary and also describes a rotation – this is easy to see intuitively as well. In particular, any rotation preserves the vector component along one axis, the axis of rotation, while it rotates the orthogonal component about that axis. Hence, only three quantities are required to describe an *arbitrary* rotation – two quantities to describe the latitude and longitude of the invariant or rotation axis and one to describe the amount of rotation around that axis. Various methods exist for describing these three parameters. For example, the Euler angles – consult the text by Goldstein *et al.* (2002) – can describe any rotation by employing three consecutive rotations, by angles α, β, and γ, made, respectively, about the z-axis, the y-axis, and the z-axis again. This result seems less intuitive, given the orthonormal decomposition available for the rows and columns of a unitary matrix. However, we begin with nine independent quantities, and observe that equations (1.51) describe six *independent* conditions, which leaves us with three independent quantities. A simple illustration of a rotation about, say, the z-axis by an angle θ is given by the matrix

$$\mathcal{A} = \begin{pmatrix} \cos\theta & \sin\theta & 0 \\ -\sin\theta & \cos\theta & 0 \\ 0 & 0 & 1 \end{pmatrix}. \quad (1.52)$$

We shall return shortly to the question of how we establish these for an arbitrary rotation matrix $\mathcal{A}$.

Recall equation (1.48) for the transformation of a vector; in indicial form this can be expressed

$$u'_i = A_{ij} u_j. \quad (1.53)$$

Similarly, a second-rank tensor **T** transforms according to

$$T'_{ij} = A_{ik} A_{jl} T_{kl}; \quad (1.54)$$

this result can be generalized readily to higher rank tensors. Moreover, we can express the inverse results

$$u_i = A_{ji} u'_j, \quad (1.55)$$

and

$$T_{ij} = A_{ki} A_{lj} T'_{kl}. \quad (1.56)$$

The existence of these transformations is what makes a tensor a very special form of matrix. This is the subject of the rotation group and is of fundamental importance in areas ranging from high energy physics to quantum chemistry.

Finally, it is important to consider the so-called *similarity* transformation. Two matrices $\mathcal{B}$ and $\mathcal{B}'$ are considered similar if there exists a rotation matrix $\mathcal{A}$ such that we can write

$$\mathcal{B}' = \mathcal{A}^T \mathcal{B} \mathcal{A}. \tag{1.57}$$

It is easy to show that the trace of the transformed matrix is identical to that of the original matrix. Suppose now that $\mathcal{B}$ is itself a rotation matrix. Then, imagine selecting $\mathcal{A}$ such that we are transforming to a new coordinate system where the rotation axis of $\mathcal{B}$ is aligned along the z-axis. Then, we observe from (1.48) that the trace must equal $1 + 2\cos\theta$, where θ is the angle of rotation. Thus, we have a simple and direct way to calculate the cosine of the angle of rotation. To calculate more, it is necessary to consider some of the properties of eigenvalues and eigenvectors of matrices.

1.4 Eigenvalues and eigenvectors

If for a given square matrix $\mathcal{A}$ one can find a number λ and a vector $\mathbf{u}$ such that the equation $\mathbf{A} \cdot \mathbf{u} = \lambda \mathbf{u}$ is satisfied, λ is said to be an *eigenvalue* or *characteristic value* and $\mathbf{u}$ an *eigenvector* or *characteristic vector* of the matrix $\mathcal{A}$. The equation can also be written

$$(\mathbf{A} - \lambda \mathbf{I}) \mathbf{u} = 0. \tag{1.58}$$

This is a linear homogeneous system of equations, and a nontrivial solution exists only if the condition $\det(\mathbf{A} - \lambda \mathbf{I}) = 0$ is met. Thus, we obtain an algebraic, i.e. polynomial, equation of degree N in λ, namely

$$\det \begin{pmatrix} A_{11} - \lambda & A_{12} & \ldots & A_{1N} \\ A_{21} & A_{22} - \lambda & \ldots & A_{2N} \\ \vdots & \vdots & \ddots & \vdots \\ A_{N1} & A_{N2} & \ldots & A_{NN} - \lambda \end{pmatrix} = 0. \tag{1.59}$$

After substantial manipulation, the latter equation can be written

$$\lambda^N - (A_{11} + A_{22} + \cdots + A_{NN}) \lambda^{N-1} + \cdots + (-1)^N \det \mathcal{A} = 0, \tag{1.60}$$

which is sometimes referred to as the *characteristic* or *eigenvalue* equation. Thus, we obtain that

$$\sum_{i=1}^{N} \lambda_i = \operatorname{tr} \mathcal{A}; \quad \text{and} \quad \prod_{i=1}^{N} \lambda_i = \det \mathcal{A}. \tag{1.61}$$

If the equation $\mathcal{A} \mathbf{u} = \lambda \mathbf{u}$ is premultiplied by $\mathcal{A}$, we get $\mathcal{A}^2 \mathbf{u} = \lambda^2 \mathbf{u}$ and, analogously, $\mathcal{A}^m \mathbf{u} = \lambda^m \mathbf{u}$ for any $m \geq 0$. Hence, the eigenvalues of $\mathcal{A}$, namely

$\lambda_1, \ldots, \lambda_N$ are transformed into the eigenvalues $\lambda_1^m, \ldots, \lambda_N^m$ of $\mathcal{A}^m$, while the eigenvectors $\mathbf{u}_1, \ldots, \mathbf{u}_N$ remain the same.

Consider now what happens under a coordinate transformation, i.e. a rotation $\mathcal{S}$, such that the new coordinates are denoted by $\mathbf{x}'$ and $\mathbf{y}'$ and the old coordinates are denoted by $\mathbf{x}$ and $\mathbf{y}$. We will employ the transformation rule

$$\mathbf{x}' = \mathcal{S}\,\mathbf{x}; \qquad \mathbf{y}' = \mathcal{S}\,\mathbf{y}. \tag{1.62}$$

Then the mapping $\mathbf{y} = \mathcal{A}\,\mathbf{x}$ becomes

$$\mathbf{y}' = \mathcal{S}\,\mathcal{A}\,\mathbf{x} = \mathcal{S}\,\mathcal{A}\,\mathcal{S}^{-1}\,\mathbf{x}' \text{ or } \mathbf{y}' = \mathcal{B}\,\mathbf{x}' \text{ with } \mathcal{B} = \mathcal{S}\,\mathcal{A}\,\mathcal{S}^{-1} \text{ or } \mathcal{A} = \mathcal{S}^{-1}\,\mathcal{B}\,\mathcal{S}; \tag{1.63}$$

since $\mathcal{S}$ describes a coordinate transformation, $\mathcal{S}^{-1} = \mathcal{S}^T$. Thus, if $\mathbf{u}$ is an eigenvector of $\mathcal{A}$ corresponding to the eigenvalue λ, then it follows that $\mathbf{v}$ is an eigenvector of $\mathcal{B}$ with the same eigenvalue λ and where

$$\mathbf{v} = \mathcal{S}^{-1}\,\mathbf{u}. \tag{1.64}$$

This result shows that the eigenvalues of a matrix are unchanged under a similarity transformation.

The *fundamental theorem of algebra* (Dummit and Foote, 2004) states that any polynomial of degree N has exactly N roots. Moreover, these roots may be complex-valued and need not be distinct; the polynomial may have multiple roots. Hence, the characteristic equation (1.60) has N roots. However, the possible complexity of roots does not always permit the existence of real-valued eigenvectors – i.e. a complex eigenvalue implies a complex eigenvector. We will show in the next chapter that, if the matrix $\mathcal{A}$ is symmetric (or Hermitian), then all of its eigenvalues are real. In the case of real-valued matrices $\mathcal{A}$, this further implies that a complete set of eigenvectors exist, although they might not be entirely unique. In cases where, for example, two eigenvalues are numerically the same, i.e. degenerate, their corresponding eigenvectors come from a two-dimensional subspace – the two eigenvectors are orthogonal and exist within that subspace, but nothing further can be said about them.

While the characteristic equation can be established from the determinant condition (1.59), it is useful to identify a direct relationship between the eigenvalues λ_i, $i = 1, \ldots, N$, and the trace of the matrix $\mathcal{A}$ as well as the next $N - 1$ powers of that matrix, a relationship often attributed to Newton. This result was generalized by the astronomer Leverrier in 1840 during his investigations of the orbit of Uranus. Suppose we express the determinant condition in the usual form

$$f(\lambda) = \lambda^N + a_1 \lambda^{N-1} + a_2 \lambda^{N-2} + \cdots + a_N = 0. \tag{1.65}$$

Exploiting the fundamental theorem of algebra, we can express this as

$$f(\lambda) \equiv (\lambda - \lambda_1)(\lambda - \lambda_2) \cdots (\lambda - \lambda_N) \tag{1.66}$$

so that we can write

$$\frac{df(\lambda)}{d\lambda} = \frac{f(\lambda)}{\lambda - \lambda_1} + \frac{f(\lambda)}{\lambda - \lambda_2} + \cdots + \frac{f(\lambda)}{\lambda - \lambda_N}, \tag{1.67}$$

where again the λ_i, for $i = 1, \ldots, N$, are the roots of the characteristic equation (1.59). However, we can also write the quotient of $f(\lambda)/(\lambda - \lambda_i)$, for $i = 1, \ldots, N$, as

$$\frac{f(\lambda)}{\lambda - \lambda_i} = \lambda^{N-1} + (\lambda + a_1)\lambda^{N-2} + (\lambda^2 + a_1\lambda + a_2)\lambda^{N-3} + \cdots, \tag{1.68}$$

by dividing the characteristic polynomial by $\lambda - \lambda_i$. We now sum these latter equations setting $\lambda = \lambda_1, \lambda_2, \ldots, \lambda_N$, respectively, and insert them in the preceding expression for $df(\lambda)/d\lambda$. For convenience, we will define S_r as

$$S_r \equiv \sum_{i=1}^{N} \lambda_i^r. \tag{1.69}$$

We can identify these quantities with the trace of $\mathcal{A}$ and its powers, i.e. $S_r = \operatorname{tr} \mathcal{A}^r$. We recall the usual polynomial form (1.65)

$$df(\lambda)/d\lambda = N\lambda^{N-1} + (N-1)a_1\lambda^{N-2} + \cdots \tag{1.70}$$

and, equating the two expressions for $df(\lambda)/d\lambda$, we then obtain

$$0 = S_1 + a_1 \tag{1.71}$$

$$0 = S_2 + a_1 S_1 + 2a_2 \tag{1.72}$$

$$\vdots$$

$$0 = S_{N-1} + a_1 S_{N-1} + \cdots + N a_N. \tag{1.73}$$

Finally, we form an expression for $f(\lambda_1) + f(\lambda_2) + \cdots + f(\lambda_N) = 0$, whereupon we obtain that

$$0 = S_N + a_1 S_{N-1} + \cdots + a_{n-1} S_1 + N a_N \tag{1.74}$$

which, together with the previous expressions, form the well-known *Newton identities*. Importantly, they provide a convenient vehicle for obtaining the characteristic polynomial by evaluating the trace of the first N powers of the matrix $\mathcal{A}$ instead of algebraically expanding the associated determinant in (1.59).

It is possible to construct *matrix functions*. For example, consider the polynomial

$$f(z) = z^n + a_1 z^{n-1} + \cdots + a_n, \tag{1.75}$$

1.4 Eigenvalues and eigenvectors

where the a_i are real or possibly complex constants. Then, replacing z by the square matrix $\mathcal{A}$, we get a matrix polynomial

$$f(\mathcal{A}) = \mathcal{A}^n + a_1 \mathcal{A}^{n-1} + \cdots + a_n \mathcal{I}. \tag{1.76}$$

This can be generalized to an infinite power series; if the usual polynomial is convergent in the domain $|z| < R$, then it can be shown that the matrix power series is also convergent so long as the magnitude of all of its eigenvalues is less than R. Two particular examples include the exponential series, which converges for *all* finite matrices, namely

$$\exp(\mathcal{A}) = \mathcal{I} + \mathcal{A} + \frac{\mathcal{A}^2}{2!} + \frac{\mathcal{A}^3}{3!} + \cdots, \tag{1.77}$$

while the series $(\mathcal{I} - \mathcal{A})^{-1} = \mathcal{I} + \mathcal{A} + \mathcal{A}^2 + \mathcal{A}^3 + \cdots$ converges only if all of the eigenvalues of $\mathcal{A}$ are less than one in magnitude.

A particularly important result in linear algebra is the *Cayley–Hamilton theorem* which states that any square matrix $\mathcal{A}$ satisfies its own characteristic equation. We will not prove this result in its general case, but we outline the proof for symmetric real-valued matrices $\mathcal{A}$. This is particularly important as symmetry is a hallmark of the stress and strain tensors and is an outcome of Newton's laws of motion. We note that one can "diagonalize" the matrix, i.e. obtain the form $\mathcal{D} = \mathcal{S}^{-1} \mathcal{A} \mathcal{S}$, where $\mathcal{S}$ is composed of column vectors corresponding to the N distinct eigenvectors $\mathbf{u}_i$, $i = 1, \ldots, N$. In this case, it immediately follows that $\mathcal{A}^n = \mathcal{S} \mathcal{D}^n \mathcal{S}^{-1}$ for $n \geq 0$; this "telescoping" property allows us to directly validate this claim. Importantly, the Cayley–Hamilton theorem tells us that $\mathcal{A}^N$, where $\mathcal{A}$ is an $N \times N$ matrix, can be expressed in terms of a linear combination of $\mathcal{A}$ taken to lower powers, as is evident from (1.76). Similarly, we can show by induction that any term of the form $\mathcal{A}^n$, for $n > N$, can be written as a linear combination of $\mathcal{A}$ taken to powers lower than N. In many situations in continuum mechanics, the matrices involved are positive semi-definite. This usually emerges in cases where we are examining some mass- or energy-like quantity. (We employed the term semi-definite to allow for situations where an associated eigenvalue may vanish but cannot become negative.) By employing the diagonal representation $\mathcal{A} = \mathcal{S} \mathcal{D} \mathcal{S}^{-1}$ and the statement that $\mathbf{x}^T \mathcal{A} \mathbf{x} \geq 0$, it follows that

$$\sum_{i=1}^{N} D_{ii} x_i^2 \geq 0, \tag{1.78}$$

where we are explicitly summing over the i index. In order for the inequality to always be satisfied, it is necessary that all diagonal components $D_{ii} = \lambda_i$ (no summation intended) be positive semi-definite. The largest eigenvalue is sometimes called the "principal value" and its corresponding eigenvector is called the "principal vector." In many situations, the principal value (sometimes called the principal component or factor) dominates the practical evolution of a problem, and there is a considerable literature dealing with its computation. We now have the necessary tools to investigate the general problem of a rotation.

Earlier in this chapter, we established that an arbitrary tensor could be additively decomposed into the sum of symmetric and anti-symmetric parts. Another important result is that any invertible tensor can be multiplicatively decomposed into the product of a rotation matrix with a symmetric, positive-definite tensor or, alternatively, the product of a positive-definite symmetric tensor with a rotation matrix. It is important to remember that a rotation matrix is itself a tensor. Thus, we say that an arbitrary invertible tensor **A** can be expressed by the forms

$$\mathbf{A} = \mathbf{Q} \cdot \mathbf{U} = \mathbf{V} \cdot \mathbf{Q}, \tag{1.79}$$

where **Q** is a rotation matrix while **U** and **V** are positive-definite symmetric tensors. We briefly summarize how to prove this result.

To show this, we need to show that $\mathbf{A}^T \cdot \mathbf{A}$ and $\mathbf{A} \cdot \mathbf{A}^T$ are positive definite and symmetric. Accordingly, we can then construct the square roots of these tensors, namely **U** and **V**, such that $\mathbf{U} \cdot \mathbf{U} = \mathbf{A}^T \cdot \mathbf{A}$ and $\mathbf{V} \cdot \mathbf{V} = \mathbf{A} \cdot \mathbf{A}^T$. Upon doing this, we consider the tensors defined by $\mathbf{A} \cdot \mathbf{U}^{-1}$ and $\mathbf{V}^{-1} \cdot \mathbf{A}$ and show in this case that these tensors are rotation matrices. Finally, we need to show that the two rotation matrices are identical and, thereby, verify the polar decompositions (1.79). Details of this derivation may be found in Chadwick (1999) or Narasimhan (1993). We turn our attention now to the nature of rotation.

1.5 Simplified approach to rotation

Suppose we have a vector **r** which we wish to rotate around a unit vector $\hat{\mathbf{u}}$ (in the sense of the right-hand rule) by an angle θ. Under this rotation, we know that the projection of **r** on $\hat{\mathbf{u}}$, namely the dyad expression $\hat{\mathbf{u}}\hat{\mathbf{u}} \cdot \mathbf{r}$, remains unchanged. Using vectors, it follows for a differential rotation that

$$d\mathbf{r} = \hat{\mathbf{u}} \times \mathbf{r}\, d\theta, \tag{1.80}$$

or in tensor form

$$d\mathbf{r} = \mathbf{U}\mathbf{r}\, d\theta, \tag{1.81}$$

where **U** is defined according to

$$U_{ij} = \epsilon_{ikj}\, \hat{u}_k. \tag{1.82}$$

1.5 Simplified approach to rotation

The solution to this differential equation can be formally written using the exponential function, i.e.

$$\mathbf{r}(\theta) = \exp(\mathbf{U}\theta)\,\mathbf{r}(0). \tag{1.83}$$

Before employing the exponential expansion (1.77), we observe that

$$U_{ij}^2 = \hat{u}_i\,\hat{u}_j - \delta_{ij}, \tag{1.84}$$

and that

$$U_{ij}^3 = -U_{ij}, \tag{1.85}$$

which we can regard as an outcome of the Cayley–Hamilton theorem. Here, we have employed the ij components of the matrices $\mathbf{U}^2$ and $\mathbf{U}^3$. Using these expressions, we can eliminate all terms emerging from the series (1.77) that are higher than the second power in $\mathbf{U}$ and obtain

$$\mathbf{r}(\theta) = \left[\mathcal{I} + \mathcal{U}\sin\theta + (1-\cos\theta)\,\mathcal{U}^2\right]\mathbf{r}(0) \tag{1.86}$$

by expanding the series and collecting terms in $\mathcal{I}$, $\mathcal{U}$, and $\mathcal{U}^2$. Hence, the rotation matrix $\mathcal{A}(\theta)$ can be written

$$A_{ij} = \delta_{ij} + \epsilon_{ijk}\,\hat{u}_k\,\sin\theta + (1-\cos\theta)\left[\hat{u}_i\,\hat{u}_j - \delta_{ij}\right]. \tag{1.87}$$

Intuitively, we recognize that a rotation of a vector preserves its projection around the $\hat{\mathbf{u}}$ axis and performs a rotation through an angle θ in the plane formed by the $\hat{\mathbf{u}} \times \mathbf{r}$ and $\hat{\mathbf{u}} \times (\hat{\mathbf{u}} \times \mathbf{r})$ axes. Accordingly, we can recognize the second and third terms on the right-hand side as being the outcomes of rotation in the plane established by the $\hat{\mathbf{u}} \times \mathbf{r}$ and $\hat{\mathbf{u}} \times (\hat{\mathbf{u}} \times \mathbf{r})$ directions, respectively. Meanwhile, it is easy to show that this rotation preserves the projection of $\mathbf{r}$ along $\hat{\mathbf{u}}$ and, furthermore, we note that the trace of this matrix is $1 + 2\cos\theta$. This conforms with our expectation for a simple two-dimensional rotation (1.52) added to the contribution to the trace emerging from the third dimension. The remaining question, how do we identify the axis of rotation from a generalized rotation matrix, can now be resolved by inspecting the off-diagonal terms, namely $\epsilon_{ijk}\,u_k\,\sin\theta$. Apart from a $\sin\theta$ normalization, they can be "read-off" the antisymmetric part of the matrix.

As a simple example, consider

$$\mathbf{A} = \begin{pmatrix} 2/3 & 2/3 & -1/3 \\ -1/3 & 2/3 & 2/3 \\ 2/3 & -1/3 & 2/3 \end{pmatrix}. \tag{1.88}$$

It is easy to show that the product of $\mathbf{A}$ with $\mathbf{A}^T$, in either order, is the identity matrix confirming that this is a rotation matrix. Since its trace is 2, it follows that $\cos\theta = 1/2$ making $\sin\theta = \sqrt{3}/2$. (Had we taken the negative root in the latter, our estimate for **u** would be reversed in sign.) Proceeding to calculate the anti-symmetric part of this particular matrix, we immediately observe that $\left(\hat{u}_1 \sin\theta, -\hat{u}_2 \sin\theta, \hat{u}_3 \sin\theta\right) = (1/2, -1/2, 1/2)$ thereby yielding $\hat{u}_1 = \hat{u}_2 = \hat{u}_3 = \sqrt{3}/3$.

1.6 Curvature, torsion, and kinematics

In order to better appreciate the complexity introduced by curvilinear coordinates, let us consider a situation where the use of a time-variable vector can be employed to better understand a dynamical problem, i.e. a vector that varies in time (or, in continuum environments, with respect to space). Once we become more deeply acquainted with continuum descriptions, we will encounter vectors that are a function of position; these are the objects that can cause substantial difficulty if they are allowed to define local coordinate systems. Suppose that we have a position **r** in two dimensions; it is useful to describe this vector by decomposing it into its magnitude r and its (unit) direction $\hat{\mathbf{u}}$ as a function of time t, namely

$$\mathbf{r}(t) = r(t)\,\hat{\mathbf{u}}(t). \tag{1.89}$$

We now wish to calculate the velocity, namely the time derivative of the position, according to

$$\frac{d\mathbf{r}(t)}{dt} = \dot{r}(t)\,\hat{\mathbf{u}}(t) + r(t)\,\frac{d\hat{\mathbf{u}}(t)}{dt}. \tag{1.90}$$

The calculation of the time derivative of the unit vector complicates the problem in a fundamental way. To better understand this, let us consider the vector $\hat{\mathbf{u}}$ at times t and $t + \Delta t$, where Δt is presumed to be "small."

Using figure 1.2 and standard definitions from calculus, it follows that

$$\frac{d\hat{\mathbf{u}}(t)}{dt} = \lim_{\Delta t \to 0} \frac{\hat{\mathbf{u}}(t + \Delta t) - \hat{\mathbf{u}}(t)}{\Delta t}. \tag{1.91}$$

Examining the figure, it follows that the difference between the two unit vectors in the numerator is the angular arc $\Delta\theta$ times a unit vector in the direction tangent to the arc. In the limit $\Delta\theta \to 0$, it follows that this unit vector is normal to $\hat{\mathbf{u}}(t)$ and we designate it as $\hat{\mathbf{n}}(t)$. Further, it is natural to express $\Delta\theta/\Delta t$ as an angular velocity, say ω, defined by

$$\lim_{\Delta t \to 0} \frac{\Delta\theta}{\Delta t} = \omega(t). \tag{1.92}$$

Thus, it follows that

$$\frac{d\hat{\mathbf{u}}(t)}{dt} = \lim_{\Delta t \to 0} \frac{\hat{\mathbf{u}}(t + \Delta t) - \hat{\mathbf{u}}(t)}{\Delta t} = \omega(t)\,\hat{\mathbf{n}}(t), \tag{1.93}$$

1.6 Curvature, torsion, and kinematics 19

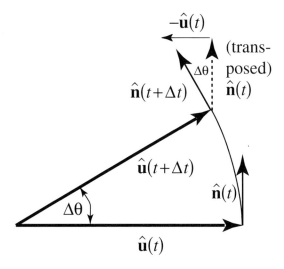

Figure 1.2 Geometry for problem of calculating acceleration.

and, consequently,

$$\frac{d\mathbf{r}(t)}{dt} = \dot{r}(t)\,\hat{\mathbf{u}}(t) + r(t)\,\omega(t)\,\hat{\mathbf{n}}(t). \tag{1.94}$$

We recognize this expression as a form of polar coordinate decomposition for the velocity vector and we have observed the complexity introduced by the time-variable character of the unit vector $\hat{\mathbf{u}}(t)$.

Now, suppose we wish to calculate as well the second derivative of $\mathbf{r}(t)$, namely the acceleration. Using the results already obtained, it is easy to show that (where the time-dependence is no longer explicitly shown)

$$\frac{d^2\mathbf{r}}{dt^2} = \ddot{r}\,\hat{\mathbf{u}} + 2\dot{r}\,\omega\,\hat{\mathbf{n}} + r\,\dot{\omega}\,\hat{\mathbf{n}} + r\,\omega\,\frac{d\hat{\mathbf{n}}}{dt}. \tag{1.95}$$

As before, it follows that

$$\frac{d\hat{\mathbf{n}}(t)}{dt} = \lim_{\Delta t \to 0} \frac{\hat{\mathbf{n}}(t + \Delta t) - \hat{\mathbf{n}}(t)}{\Delta t}. \tag{1.96}$$

Recalling that $\hat{\mathbf{n}}(t)$ is strictly orthogonal to $\hat{\mathbf{u}}(t)$, it follows from figure 1.2 that the same arc $\Delta\theta$ in magnitude is swept out by $\hat{\mathbf{n}}(t)$ and that the direction in which this occurs is given by $-\hat{\mathbf{u}}(t)$. Hence, it follows that

$$\frac{d\hat{\mathbf{n}}(t)}{dt} = \lim_{\Delta t \to 0} \frac{\hat{\mathbf{n}}(t + \Delta t) - \hat{\mathbf{n}}(t)}{\Delta t} = -\omega(t)\,\hat{\mathbf{u}}(t), \tag{1.97}$$

so that
$$\frac{d^2\mathbf{r}}{dt^2} = \left[\ddot{r} - r\,\omega^2\right]\hat{\mathbf{u}} + \left[2\dot{r}\,\omega + r\,\dot{\omega}\right]\hat{\mathbf{n}}. \tag{1.98}$$

The latter term can be expressed in terms of the angular momentum, that is
$$2\dot{r}\,\omega + r\,\dot{\omega} = \frac{1}{r}\frac{d}{dt}\left(r^2\omega\right). \tag{1.99}$$

When there is no non-radial acceleration, then the angular momentum is preserved. This is also apparent when we take the radial cross-product of $\mathbf{r}\times$ with the former equation to obtain the torque. Meanwhile in the radial direction, we observe that the acceleration in the radial direction is reduced according to the square of the angular velocity. This is the familiar centripetal acceleration term. The non-radial term, however, is more complex and we observe that there is an additional term $2\dot{r}\,\omega$ which is associated with the *Coriolis force*. This emerges when the rotation ω is associated with the coordinates that are themselves undergoing a rotation thereby resulting in non-inertial effects. Thus, we see how time-variable as well as space-variable vector geometry complicates these problems. For this reason, we adhere to Cartesian coordinates in our derivations, introducing curvilinear coordinates only when the associated vector equations have been derived.

Finally, to conclude this discussion, let us assume that the motion is *not* confined to two dimensions. It should be pointed out that in our previous discussion we assumed that $\hat{\mathbf{n}}(t)$ was in the plane defined by $\hat{\mathbf{u}}(t)$ and $\hat{\mathbf{u}}(t + \Delta t)$; however, $\hat{\mathbf{n}}(t + \Delta t)$ is in the plane defined by $\hat{\mathbf{u}}(t + \Delta t)$ and $\hat{\mathbf{u}}(t + 2\Delta t)$ and we can visualize situations where $\hat{\mathbf{u}}(t + 2\Delta t)$ is *not* in the plane defined by $\hat{\mathbf{u}}(t)$ and $\hat{\mathbf{u}}(t + \Delta t)$. This would occur if the plane of the trajectory is constantly changing. We would expect this additional complication to emerge if our trajectory has a helical character or, in the language of differential geometry, shows *torsion*. This issue is generally described under the topic of the *Serret–Frenet formulae*; see, for example, Mathews and Walker (1970) or Schutz (1980) or Aris (1989). This topic is mathematically very rich, and deeper mathematical treatments can be found in Millman and Parker (1977) and in Carmo (1976). Our derivation, however, will be somewhat different in order to provide a connection with dynamics.

Recall that $\hat{\mathbf{u}}(t)$ is a unit vector. Hence, it follows that
$$\frac{1}{2}\frac{d}{dt}\left|\hat{\mathbf{u}}(t)\right|^2 = \hat{\mathbf{u}}(t) \cdot \frac{d\hat{\mathbf{u}}(t)}{dt} = 0, \tag{1.100}$$

thereby establishing mathematically that $\hat{\mathbf{u}}(t)$ and $\dot{\hat{\mathbf{u}}}(t)$ are orthogonal, as we observed earlier in figure 1.2 from geometrical considerations. Accordingly, we define $\omega(t)$ according to
$$\frac{d\hat{\mathbf{u}}(t)}{dt} \equiv \omega(t)\,\hat{\mathbf{n}}(t), \tag{1.101}$$

1.6 Curvature, torsion, and kinematics 21

where $\hat{\mathbf{n}}(t)$ is the unit normal vector identified earlier (sometimes called the principal normal), and $\omega(t)$ is the angular velocity.

Let us now relax the assumption made earlier that $\dot{\hat{\mathbf{n}}}(t)$ is in the plane defined by $\hat{\mathbf{u}}(t)$ and $\hat{\mathbf{n}}(t)$. For reasons analogous to those above, it follows that

$$\hat{\mathbf{n}}(t) \cdot \dot{\hat{\mathbf{n}}}(t) = 0 \tag{1.102}$$

and, therefore, we can write

$$\frac{d\hat{\mathbf{n}}(t)}{dt} = C_n(t)\,\hat{\mathbf{u}}(t) + C_w(t)\,\hat{\mathbf{w}}(t), \tag{1.103}$$

where we define $\hat{\mathbf{w}}$ according to

$$\hat{\mathbf{w}}(t) \equiv \hat{\mathbf{u}}(t) \times \hat{\mathbf{n}}(t). \tag{1.104}$$

Note that $\hat{\mathbf{u}}$, $\hat{\mathbf{n}}$, and $\hat{\mathbf{w}}$ form a right-handed coordinate system at every instant in time. Recalling that $\hat{\mathbf{u}}(t) \cdot \hat{\mathbf{n}}(t) = 0$, we observe upon taking its time derivative that

$$\hat{\mathbf{u}}(t) \cdot \frac{d\hat{\mathbf{n}}(t)}{dt} + \frac{d\hat{\mathbf{u}}(t)}{dt} \cdot \hat{\mathbf{n}}(t) = 0. \tag{1.105}$$

Combining this expression with the ones for $\dot{\hat{\mathbf{u}}}$ and $\dot{\hat{\mathbf{n}}}$ yields

$$C_n(t) = -\omega(t). \tag{1.106}$$

This result corresponds to our earlier observation, valid in the planar case, but we must now establish whether $C_w(t)$ is significant. We will replace the role of $-C_w(t)$ by $\eta(t)$ which describes the "pitch" of the helical path in angular velocity terms, i.e. the angle of sweep out of the plane per unit time. To do this, let us consider the time derivative of $\hat{\mathbf{w}}$ with each member of the triad $\hat{\mathbf{u}}$, $\hat{\mathbf{n}}$, and $\hat{\mathbf{w}}$, namely

$$\begin{aligned}\frac{d(\hat{\mathbf{u}} \cdot \hat{\mathbf{w}})}{dt} &= \dot{\hat{\mathbf{u}}} \cdot \hat{\mathbf{w}} + \hat{\mathbf{u}} \cdot \dot{\hat{\mathbf{w}}} = 0 \\ \frac{d(\hat{\mathbf{n}} \cdot \hat{\mathbf{w}})}{dt} &= \dot{\hat{\mathbf{n}}} \cdot \hat{\mathbf{w}} + \hat{\mathbf{n}} \cdot \dot{\hat{\mathbf{w}}} = 0 \\ \frac{d(\hat{\mathbf{w}} \cdot \hat{\mathbf{w}})}{dt} &= 2\hat{\mathbf{w}} \cdot \dot{\hat{\mathbf{w}}} = 0.\end{aligned} \tag{1.107}$$

From the first and third of these equations, it follows that $\dot{\hat{\mathbf{w}}}$ is orthogonal to $\hat{\mathbf{u}}$ and to $\hat{\mathbf{w}}$, thereby putting it solely in the $\mathbf{n}$ direction. The second equation then yields

$$\eta(t) = \hat{\mathbf{n}}(t) \cdot \dot{\hat{\mathbf{w}}}(t). \tag{1.108}$$

Thus, we conclude with a system of three dynamical equations

$$\frac{d\hat{\mathbf{u}}(t)}{dt} = \omega(t)\,\hat{\mathbf{n}}(t)$$
$$\frac{d\hat{\mathbf{n}}(t)}{dt} = -\omega(t)\,\hat{\mathbf{u}}(t) - \eta(t)\,\hat{\mathbf{w}}(t) \qquad (1.109)$$
$$\frac{d\hat{\mathbf{w}}(t)}{dt} = \eta(t)\,\hat{\mathbf{n}}(t).$$

This description of vector behavior can be applied to environments where material is undergoing substantial shear or rotation.

Before departing from our geometrically motivated discussion of rotation, it is useful to review some aspects of coordinate transformations that are typically encountered in elementary courses. In the preceding two-dimensional discussion, we effectively employed the polar coordinates r and θ which are related to the corresponding Cartesian coordinates x and y in an elementary way, namely

$$x = r\cos\theta$$
$$y = r\sin\theta. \qquad (1.110)$$

It is important to recall how to obtain the various differential operations, particularly the Laplacian ∇^2, in different coordinates. For example, using the chain rule, we observe for any function f that

$$\frac{\partial f}{\partial x} = \frac{\partial r}{\partial x}\bigg|_y \frac{\partial f}{\partial r}\bigg|_\theta + \frac{\partial \theta}{\partial x}\bigg|_y \frac{\partial f}{\partial \theta}\bigg|_r, \qquad (1.111)$$

where we have been explicit in identifying what quantities are kept constant in each derivative, and so on. Accordingly, we find that

$$\frac{\partial f}{\partial x} = \cos\theta\,\frac{\partial f}{\partial r} - \frac{\sin\theta}{r}\,\frac{\partial f}{\partial \theta}$$
$$\frac{\partial f}{\partial y} = \sin\theta\,\frac{\partial f}{\partial r} + \frac{\cos\theta}{r}\,\frac{\partial f}{\partial \theta}. \qquad (1.112)$$

Adding the third or z dimension is straightforward and, after some algebra, one can derive that the Laplacian becomes in Cartesian and cylindrical coordinates respectively

$$\nabla^2 f = \frac{\partial^2 f}{\partial x^2} + \frac{\partial^2 f}{\partial y^2} + \frac{\partial^2 f}{\partial z^2} = \frac{1}{r}\frac{\partial}{\partial r}\left(r\frac{\partial f}{\partial r}\right) + \frac{1}{r^2}\frac{\partial^2 f}{\partial \theta^2} + \frac{\partial^2 f}{\partial z^2}. \qquad (1.113)$$

In spherical coordinates defined by

$$z = r \cos \theta$$
$$x = r \sin \theta \cos \phi \quad (1.114)$$
$$y = r \sin \theta \sin \phi,$$

we then obtain by a similar algebraic exercise the Laplacian

$$\nabla^2 f = \frac{1}{r^2} \frac{\partial}{\partial r} \left(r^2 \frac{\partial f}{\partial r} \right) + \frac{1}{r^2 \sin \theta} \frac{\partial}{\partial \theta} \left(\sin \theta \frac{\partial f}{\partial \theta} \right) + \frac{1}{r^2 \sin^2 \theta} \frac{\partial^2 f}{\partial \phi^2}. \quad (1.115)$$

While our derivations of the principles of continuum mechanics are simplified when performed in Cartesian geometry, the application of the equations derived is often simplified when they are expressed using the curvilinear coordinate system that is most relevant to the physical problem at hand, and it is always useful to review some of the underlying operator identities and their method of derivation. Now, we will return to our general discussion confining our description to the use of Cartesian coordinates.

Exercises

1.1 Prove the following identities using the Kronecker delta and Levi-Civita permutation symbol identities.

   (1) Show that

   $$(\mathbf{a} \times \mathbf{b}) \cdot \mathbf{a} = 0.$$

   (2) Show that

   $$\nabla \times (\nabla \times \mathbf{u}) = \nabla (\nabla \cdot \mathbf{u}) - \nabla^2 \mathbf{u}.$$

   (3) Suppose you have a set of vectors $\mathbf{u}_i$, for $i = 1, 2, \ldots, n$ in an n-dimensional space. Further suppose that $\det \mathcal{A} \neq 0$ where the ith row of $\mathcal{A}$ corresponds to the vector $\mathbf{u}_i$. Show that $\hat{\mathbf{v}}_i$, for $i = 1, 2, \ldots, n$, defined below constitute an orthogonal basis set (*Gram–Schmidt orthogonalization*):

   $$\hat{\mathbf{v}}_1 \equiv \frac{\mathbf{u}_1}{|\mathbf{u}_1|}$$

   and

   $$\hat{\mathbf{v}}_m \equiv \frac{\sum_{i=1}^{m-1} \hat{\mathbf{v}}_i \left(\hat{\mathbf{v}}_i \cdot \mathbf{u}_m \right) - \mathbf{u}_m}{\left| \sum_{i=1}^{m-1} \hat{\mathbf{v}}_i \left(\hat{\mathbf{v}}_i \cdot \mathbf{u}_m \right) - \mathbf{u}_m \right|} \text{ for } m = 2, \ldots, n,$$

where we have employed *dyads* in the summation. (That is, prove that $\hat{\mathbf{v}}_i \cdot \hat{\mathbf{v}}_j = 0$ if $i \neq j$.)

1.2 Suppose that **b** is an arbitrary point in 3-space. Let $\mathbb{X}$ be the set of points **x** such that $(\mathbf{x} - \mathbf{b}) \cdot \mathbf{x} = 0$. Show that this describes the surface of a sphere with center $\frac{1}{2}\mathbf{b}$ and radius $\frac{1}{2}b$.

1.3 Consider the volume integral

$$I_j = \int_\Omega \partial_i T_{ij} d^3x,$$

where Ω describes some finite volume with a surface Σ. Then, show that

$$\mathbf{I} = \int_\Sigma \hat{\mathbf{n}} \cdot \mathbf{T}\, d^2x.$$

Hint, recall Gauss' theorem and its symbolic equivalent. Similarly, show that

$$\int_\Sigma \left(\phi\, \psi_{,i} - \psi\, \phi_{,i}\right) n_i d^2x = \int_\Omega \left(\phi\, \psi_{,ii} - \psi\, \phi_{,ii}\right) d^3x,$$

where ϕ and ψ are scalar functions of the coordinates.

1.4 Show that the matrix

$$[C_{i,j}] \equiv \begin{pmatrix} 5 & 2 \\ -12 & -5 \end{pmatrix}$$

is a "square root" of the 2×2 identity matrix. What does this mean? What are the eigenvalues of $\mathcal{C}$? Its eigenvectors?

1.5 Recall that the transformation properties of a tensor give

$$T'_{ij} = A_{il}\, A_{jm}\, T_{lm},$$

where $\mathcal{A}$ describes a rotation matrix. Show that this is equivalent to

$$\mathcal{T}' = \mathcal{A}\, \mathcal{T}\, \mathcal{A}^T,$$

i.e. that rotation of a second-rank tensor is equivalent to a similarity transformation.

1.6 Prove the following identities using the Kronecker delta and permutation symbol properties, where λ and ϕ are scalar functions of the coordinates x_i.

(1) Bernoulli identity

$$(\mathbf{v} \cdot \nabla)\, \mathbf{v} = \frac{1}{2} \nabla v^2 - \mathbf{v} \times (\nabla \times \mathbf{v}).$$

(2) Divergence of scalar product

$$\nabla \cdot (\lambda \nabla \phi) = \lambda \nabla^2 \phi + \nabla \lambda \cdot \nabla \phi.$$

(3) Laplacian of scalar product

$$\nabla^2 (\lambda \phi) = \lambda \nabla^2 \phi + 2 (\nabla \lambda) \cdot (\nabla \phi) + \phi \nabla^2 \lambda.$$

1.7 For the position vector x_i having a magnitude x, we define *contravariant differentiation* according to

$$x_{,j} \equiv \partial_j x = \frac{x_j}{x}.$$

Similarly, if there appear two indices following a subscripted comma, then we differentiate with respect to each of those indices. Therefore, show the following.

(1) Mixed derivatives of length

$$x_{,ij} = \frac{\delta_{ij}}{x} - \frac{x_i x_j}{x^3}.$$

(2) Mixed derivatives of reciprocal of length

$$\left(x^{-1}\right)_{,ij} = \frac{3 x_i x_j}{x^5} - \frac{\delta_{ij}}{x^3}.$$

(3) Laplacian of length

$$x_{,ii} = \frac{2}{x}.$$

1.8 Suppose $\hat{\mathbf{u}} \equiv (0, 0, 1)$. Using the identity derived earlier for a general rotation about $\hat{\mathbf{u}}$, what is the explicit matrix form for a rotation by angle θ?

1.9 Consider two *arbitrary* unit vectors $\hat{\mathbf{v}}$ and $\hat{\mathbf{w}}$. Consider a rotation made about $\hat{\mathbf{v}}$ by an angle θ and then a rotation made about $\hat{\mathbf{w}}$ by an angle ϕ. What is the combined effect of the two rotations, i.e. what is the angle of rotation? Also, what is the angle if the order of the two operations is reversed? Are these angles the same?

1.10 Prove the polar decomposition theorem (1.79), following the steps provided in the earlier discussion.

1.11 Derive the Laplacian in cylindrical coordinates (1.113).

1.12 Derive the Laplacian in spherical coordinates (1.114).

1.13 Calculate the Laplacian for $1/r$ in spherical coordinates in two different ways. You wish to show that it vanishes for $r > 0$.

(1) Calculate the trace of the matrix derived above in problem 1.8, i.e. the trace of $\left(x^{-1}\right)_{,ij}$. Remember that x in that problem is the length of the Cartesian vector $\mathbf{x}$ making it identical to r.

(2) Employ the expression (1.114) in spherical coordinates.

1.14 For reasons that will become evident in chapter 5, we need to be able to evaluate the integral over all space of $\nabla^2 (1/r)$, namely $\int \nabla^2 (1/r) \, d^3r$. The integrand, as you have just shown, vanishes so long as $r > 0$, but there is a singularity present at $r = 0$. Evaluate this integral, showing that it is numerically equal to -4π. Hint: employ Gauss' theorem employing a sphere of arbitrarily small radius surrounding the origin.

2
Stress principles

2.1 Body and surface forces

We begin by examining the nature of forces on continuous media. We will pursue this theme later by examining material response. This topic is a truly venerable one with significant references made by Newton (Chandrasekhar, 1995) and many others before and after. Early treatments of this topic employ modes of notation very similar to ours, but largely focus on two-dimensional problems since much of the algebra reduces to that associated with quadratic equations. We will show that in three dimensions, the algebraic problem corresponds to a cubic polynomial with real roots which can be easily determined by analytical means. A medium is *homogeneous* if its properties are the same everywhere.

Homogeneity, however, can be of two types: regular or random. A regular homogeneous medium has the same underlying character everywhere, e.g. a piece of metal whose atoms are organized in a lattice. A random homogeneous medium has the same underlying *statistical* distribution of properties, but may lack regularity. For example, a rock composed of many different grains cemented together can be said to be homogeneous if the statistical properties of the mix do not vary. A homogeneous material is also said to be *isotropic*, i.e. looks the same in all directions. The material looks the same because it *is* the same. However, an isotropic material is not necessarily homogeneous. For example, the Earth appears to be (crudely) isotropic as viewed from its center but the core, mantle, and lithosphere are very distinct from each other. Similarly, layered sediments may be homogeneous but will not always be isotropic in their properties owing to the mode of deposition and lithification (i.e., formation of rock) of their constituent grains (e.g. shale). Only in situations where a medium appears isotropic everywhere can we say that it is also homogeneous.

We distinguish between properties that are *intrinsic* and *extrinsic*. By this we mean that intrinsic properties describe the local character of the medium, for example, its density, while extrinsic properties describe some bulk characterization, for example, its total mass. Sometimes, intrinsic quantities are determined by some extrinsic quantity over some arbitrarily small volume. Thus, temperature is a measure of (i.e. proportional to) internal energy per atom (or unit mass). Similarly, we define the (mass) density of a material as the mass Δm contained by some volume ΔV divided by that volume in the mathematical limit, namely

$$\rho = \lim_{\Delta V \to 0} \frac{\Delta m}{\Delta V}. \qquad (2.1)$$

This definition is clearly a mathematical idealization. In practice, we consider ΔV that is on one hand very small but sufficiently large to contain a statistically significant number of molecules so that the quotient will be well defined. The density is a scalar function of position and time as indicated by

$$\rho = \rho(x_i, t) = \rho(\mathbf{x}, t). \qquad (2.2)$$

In figure 2.1, we consider a body (shaded) having a volume V enclosed by a surface S in three-dimensional space. We denote by P an interior point located within the small element (unshaded) of volume ΔV whose mass is Δm.

We now distinguish between two different types of forces. First, we define *body forces* to be those which act on the entire volume of material and, second, *surface forces* as those which act on the surface alone. Gravity and electrostatic acceleration are the best-known examples of body forces while surface forces are associated with friction, viscosity and so on.

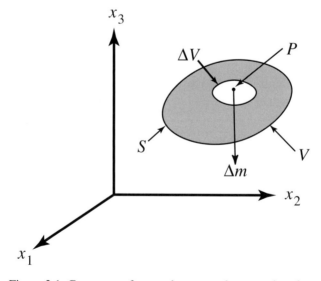

Figure 2.1 Geometry of mass element, volume, and surface.

2.2 Cauchy stress principle

Here, we do not distinguish between a bounding surface or an arbitrary element of surface within the body. (On a microscopic level, surface forces are due to very short-range forces acting between material at an interface. Mathematically, it is convenient to treat these in an idealized way.)

We designate body forces by the vector symbol b_i (force per unit mass) or p_i (force per unit volume). The two designations of body forces are related according to

$$\rho \mathbf{b} = \mathbf{p}. \tag{2.3}$$

Surface forces are designated by the vector symbol f_i and have the dimensions of a force, e.g. Newtons or dynes.

2.2 Cauchy stress principle

Suppose we now construct a plane surface, the "cutting plane" distinguished by the superscript $\star$, which we call $S^\star$ passing through P. The unit vector $\hat{n}_i$ defines the normal to the plane and a small area $\Delta S^\star$ circumscribes that point. Note that the surface force f_i need not point in the same direction as $\hat{n}_i$. This issue is of fundamental importance. Surface forces, especially, do not always act in the same direction as the normal to the surface. Friction and viscous stress are among these enigmatic yet common examples. The fact that one direction (that of the surface normal) does not translate directly into the other (that of the force) is what gives continuum mechanics its tensor character instead of a simple vector one.

The *Cauchy stress principle* defines a stress or traction vector $t_i^{(\hat{n})}$ according to

$$t_i^{(\hat{n})} = \lim_{\Delta S^\star \to 0} \frac{\Delta f_i}{\Delta S^\star}. \tag{2.4}$$

We can think of the traction as being the stress or force per unit area in the ith direction associated with a surface with normal vector $\hat{\mathbf{n}}$. It is important to note that $t_i^{(\hat{n})}$ depends on the direction of the normal vector and that, just as there are an infinity of cutting planes (and normal vectors), there are an infinity of associated stress vectors. The totality of pairs $t_i^{(\hat{n})}$ and $\hat{\mathbf{n}}$ defines the *state of stress* at that point.

Although not shown on figure 2.2, the moment vector m_i is defined according to $\epsilon_{ijk} x_j f_k$ where x_j is measured from P. In the limit $\lim_{\Delta S^\star \to 0} \Delta m_i / \Delta S^\star$ (where Δm_i denotes the infinitesimal moment associated with the infinitesimal displacement from P), we obtain a vanishing result, implying that moments (or couples) vanish on an infinitesimal surface, although there are some degenerate cases where non-zero coupled stress may be important.

In order to proceed from here, we must now introduce Newton's laws. We do so in integral form, where we integrate over an arbitrary volume contained within

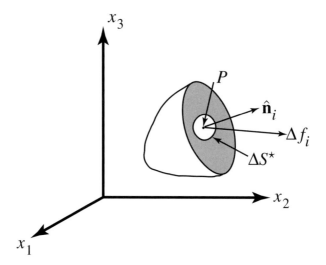

Figure 2.2 Geometry of cutting plane with normal and traction vectors.

an arbitrary surface. Suppose, we divide our volume into two arbitrary segments, denoted by Roman numerals I and II. We do not require that the separating surface be planar. Then, the force law gives us that

$$\int_{S_I} t_i^{(\hat{n})} \, d^2x + \int_{V_I} \rho \, b_i \, d^3x = \frac{d}{dt} \int_{V_I} \rho \, v_i \, d^3x, \qquad (2.5)$$

where v_i denotes the velocity of the material, and similarly

$$\int_{S_{II}} t_i^{(\hat{n})} \, d^2x + \int_{V_{II}} \rho \, b_i \, d^3x = \frac{d}{dt} \int_{V_{II}} \rho \, v_i \, d^3x. \qquad (2.6)$$

However, we could write a similar equation for the entire volume, namely

$$\int_{S} t_i^{(\hat{n})} \, d^2x + \int_{V} \rho \, b_i \, d^3x = \frac{d}{dt} \int_{V} \rho \, v_i \, d^3x. \qquad (2.7)$$

Taking the sum of the first two integral relations and subtracting from the latter, we obtain

$$\int_{S^\star} \left[t_i^{(\hat{n})} + t_i^{-(\hat{n})} \right] d^2x = 0, \qquad (2.8)$$

where $S^\star$ is the interface between the two subvolumes. The superscripts $(\hat{n})$ and $-(\hat{n})$ indicate that the relevant local normals are exactly opposite to each other. Since the latter holds for *any* surface, we obtain

$$t_i^{(\hat{n})} = -t_i^{-(\hat{n})} \qquad (2.9)$$

everywhere. This symmetry property will now be utilized in deducing the stress tensor.

2.3 Stress tensor

A point in a continuum of matter has a stress vector for each of its three coordinate planes, which can be resolved into one component normal to the plane and two tangential to it. Thus, owing to the principle of superposition, a tensor description is *complete*. The three stress vectors associated with the coordinate planes can be written in the form

$$\mathbf{t}^{(\hat{\mathbf{e}}_i)} = t_1(\hat{\mathbf{e}}_i)\,\hat{\mathbf{e}}_1 + t_2(\hat{\mathbf{e}}_i)\,\hat{\mathbf{e}}_2 + t_3(\hat{\mathbf{e}}_i)\,\hat{\mathbf{e}}_3 = t_j(\hat{\mathbf{e}}_i)\,\hat{\mathbf{e}}_j; \tag{2.10}$$

for each $i = 1, 2, 3$ this allows us to write

$$\mathbf{t}^{(\hat{\mathbf{n}})} = \mathbf{t}^{(\hat{\mathbf{e}}_j)}\,\hat{n}_j, \tag{2.11}$$

where we sum over the index j. Thus, $t_i^{(\hat{\mathbf{n}})} = t_i^{(\hat{\mathbf{e}}_j)} n_j$.

This expression shows us how to relate the force experienced with respect to the normal plane selected. In particular, we can rewrite the latter in the form

$$t_i^{(\hat{\mathbf{n}})} = \sigma_{ji}\,n_j, \tag{2.12}$$

or

$$\mathbf{t}^{(\hat{\mathbf{n}})} = \hat{\mathbf{n}} \cdot \mathbf{\Sigma}. \tag{2.13}$$

Here, we take

$$\mathbf{\Sigma} = \begin{pmatrix} \sigma_{11} & \sigma_{12} & \sigma_{13} \\ \sigma_{21} & \sigma_{22} & \sigma_{23} \\ \sigma_{31} & \sigma_{32} & \sigma_{33} \end{pmatrix}. \tag{2.14}$$

Further, we can now make the identification

$$\sigma_{ji} \equiv t_i^{(\hat{\mathbf{e}}_j)}. \tag{2.15}$$

These equations define the *Cauchy stress tensor*. Conceptually, this quantity describes the stress in the ith direction due to a surface component in the jth direction. Its geometrical meaning can be inferred from the accompanying figure.

Here we see that we associate with each coordinate plane – corresponding to a normal vector aligned with a particular coordinate axis – three stress components. We refer to the stress components aligned with the normal vectors associated with each of the planes as *normal stresses*, namely σ_{11}, σ_{22}, and σ_{33}. Similarly, we refer to the other stress components which are orthogonal to the normal stresses as *shear*

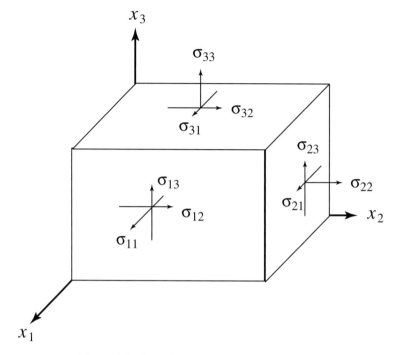

Figure 2.3 Cartesian geometry of stress tensor.

stresses, namely σ_{12}, σ_{13}, σ_{21}, σ_{23}, σ_{31}, and σ_{32}. We refer to a stress component as being positive if its vector arrow in the diagram (figure 2.3) points in the same direction as the associated coordinate axis.

Positive normal stresses are referred to as *tensile stresses*, while negative normal stresses are referred to as *compressive stresses*. Stress has the dimensionality of a pressure or force per unit area. In chapter 7, we will introduce concepts of dimensional analysis during our discussion of geophysical fluid dynamics, and we will show that all physical quantities can be identified with the dimensions of mass M, length L, and time T. Accordingly, we will identify there that stress has the dimensionality of $[ML^{-1}T^{-2}]$. The standard unit of stress in the SI system is measured as Newtons per square meter where one N/m² is defined to be one Pascal and abbreviated as Pa. In cgs units, we employ dynes per square centimeter—1 Pascal $\equiv$ 10 dynes per square centimeter. Another unit frequently encountered is the bar (for one barometric unit of atmospheric pressure): 1 bar $\approx 1.01 \times 10^5$ Pa. A typical pressure seen in a planetary core is the order of one megabar. Similarly, the stress drop experienced during an earthquake is typically 20 bar which is equivalent to 2 megapascals.

2.4 Symmetry and transformation laws

Physically, the net force on an element of matter depends on the *difference* in the surface forces acting across the element or, as the element shrinks, on the "gradients of stress." Elaborating on Newton's third law of motion, excluding for the moment the role of acceleration, we consider a material body as having a volume V and a bounding surface S and subject the body to surface tractions $t_i^{(\hat{n})}$ and body forces b_i. The condition requiring that all net forces be zero has the integral form

$$\int_S t_i^{(\hat{n})} \, d^2x + \int_V \rho \, b_i \, d^3x = 0. \tag{2.16}$$

Since $t_i^{(\hat{n})} = \sigma_{ji} \hat{n}_j$, we can employ Gauss' theorem to obtain

$$\int_S \sigma_{ji} \hat{n}_j \, d^2x = \int_V \sigma_{ji,j} \, d^3x, \tag{2.17}$$

so that (2.16) becomes

$$\int_V \left(\sigma_{ji,j} + \rho \, b_i \right) d^3x = 0. \tag{2.18}$$

Since this expression must hold for any arbitrary volume V, we obtain the *local equilibrium conditions*

$$\sigma_{ji,j} + \rho \, b_i = 0. \tag{2.19}$$

We will now explore how Newton's "action–reaction" principle results in the stress tensor being symmetric.

In addition to this force balance requirement, we must also impose a moment balance requirement. Physically, this is equivalent to saying that there is a zero net torque. Employing the same kind of volume integral decomposition, we obtain

$$\int_S \epsilon_{ijk} \, x_j \, t_k^{(\hat{n})} \, d^2x + \int_V \epsilon_{ijk} \, x_j \, \rho \, b_k \, d^3x = 0 \tag{2.20}$$

and now make use of $t_k^{(\hat{n})} = \sigma_{qk} \hat{n}_q$. Gauss' theorem then yields

$$\int_V \epsilon_{ijk} \left[\left(x_j \, \sigma_{qk} \right)_{,q} + x_j \, \rho \, b_k \right] d^3x = 0, \tag{2.21}$$

which can now be written

$$\int_V \epsilon_{ijk} \left[x_{j,q} \, \sigma_{qk} + x_j \left(\sigma_{qk,q} + \rho \, b_k \right) \right] d^3x = 0. \tag{2.22}$$

Since $x_{j,q} = \delta_{jq}$ and using (2.19), the latter equation reduces to

$$\int_V \epsilon_{ijk} \, \sigma_{jk} \, d^3x = 0, \tag{2.23}$$

whence we conclude that the integrand vanishes for all i. Thus, we have

$$\sigma_{jk} = \sigma_{kj}. \tag{2.24}$$

Given this result, it is more convenient to write (2.19) as

$$\sigma_{ij,j} + \rho\, b_i = 0, \tag{2.25}$$

or in symbolic form

$$\nabla \cdot \mathbf{\Sigma} + \rho\, \mathbf{b} = 0, \tag{2.26}$$

where $\mathbf{\Sigma}$ is the tensorial form of the stress. Similarly, we rewrite the Cauchy stress formula in the more convenient tensor form

$$t_i^{(\hat{\mathbf{n}})} = \sigma_{ij}\, \hat{n}_j, \tag{2.27}$$

or symbolic form

$$\mathbf{t}^{(\hat{\mathbf{n}})} = \mathbf{\Sigma} \cdot \hat{\mathbf{n}}. \tag{2.28}$$

Given that the stress is a tensor, because it describes Newtonian forces in any rotated coordinate system, it necessarily obeys the coordinate transformation laws

$$\sigma'_{ij} = a_{iq}\, a_{jm}\, \sigma_{qm}, \tag{2.29}$$

or

$$\boldsymbol{\sigma}' = \mathbf{A} \cdot \mathbf{\Sigma} \cdot \mathbf{A}^T. \tag{2.30}$$

We have examined in detail earlier how to obtain the rotational transformation matrix $\mathcal{A}$.

2.5 Principal stresses and directions

The *principal stresses* and *directions* are the outcome of solving the eigenvalue problem

$$\left(\sigma_{ij} - \delta_{ij}\, \sigma\right) \hat{n}_j = 0, \tag{2.31}$$

where the unsubscripted variable σ now denotes the eigenvalue or principal stress and the vector $\hat{n}_j$ denotes the eigenvector or principal direction. We have already observed, using the fundamental theorem of algebra, that there are three (possibly including multiple or complex-valued) eigenvalues for this problem – see, for example, Boas (2006). We will now show that *all* eigenvalues are real and that, therefore, all three eigenvectors exist.

Since the stress tensor $\mathbf{\Sigma}$ is real and symmetric, it is also Hermitian, i.e. $\sigma_{ij} = \sigma_{ji}^*$. This is important as it can be used to show that *all* eigenvalues of a

2.5 Principal stresses and directions

symmetric real matrix are real – a fact that can be shown by demonstrating that a logical contradiction otherwise emerges. Suppose we assume that we have two eigenvalues, denoted by λ^u and λ^v, which are complex conjugates of each other. Then, let **u** be a nontrivial (complex) solution to

$$\sigma_{ij} u_j = \lambda^u u_i; \qquad (2.32)$$

we do not refer here to **u** as an eigenvector since its assumed complex character eliminates this geometrical interpretation. Then, if we define $\mathbf{v} \equiv \mathbf{u}^*$, it follows that

$$\sigma_{ij} v_j = \lambda^v v_i. \qquad (2.33)$$

Multiplying the first of these two equations by u_i^* and the second by v_i^* and subtracting the two from each other, we obtain

$$\sigma_{ij} \left(u_i^* u_j - v_i^* v_j \right) = \sigma_{ij} \left(u_i^* u_j - u_i u_j^* \right) = 0, \qquad (2.34)$$

since $\sigma_{ij} = \sigma_{ji}^*$ and, therefore,

$$0 = \lambda^u u_i u_i^* - \lambda^v v_i v_i^* = \left(\lambda^u - \lambda^v \right) u_i u_i^*. \qquad (2.35)$$

Since $u_i u_i^* > 0$, we must conclude that $\lambda^u = \lambda^v$ contradicting our hypothesis that they are complex conjugate roots.

Having shown that the eigenvalues of a real-valued stress tensor are necessarily real, it follows that their eigenvectors exist, i.e. are themselves real valued. Suppose, now, that λ^u and λ^v are real eigenvalues of $\mathbf{\Sigma}$ and that $\hat{\mathbf{u}}$ and $\hat{\mathbf{v}}$ are their respective eigenvectors both normalized to unit length, as in (2.32) and (2.33). Multiplying the first of these equations by v_i and the second by u_i and subtracting, we observe that

$$\sigma_{ij} \left(\hat{v}_i \hat{u}_j - \hat{u}_i \hat{v}_j \right) = \sigma_{ij} \left(\hat{u}_i \hat{v}_j - \hat{u}_i \hat{v}_j \right) = 0$$
$$= \lambda^u \hat{u}_i \hat{v}_i - \lambda^v \hat{v}_i \hat{u}_i = \left(\lambda^u - \lambda^v \right) \hat{u}_i \hat{v}_i. \qquad (2.36)$$

Since the two eigenvalues were assumed to be different, it follows that $\hat{u}_i \hat{v}_i = 0$, or that their respective eigenvectors are orthogonal. Therefore, if all three eigenvalues, say λ^k, for $k = 1, 2, 3$, are distinct, the three (normalized) eigenvectors $\mathbf{u}^k$ are necessarily orthogonal and constitute a complete basis set. It follows then that the eigenvalue decomposition of $\mathbf{\Sigma}$, namely $\mathbf{\Sigma} = \mathbf{A} \cdot \mathbf{D} \cdot \mathbf{A}^T$, comes from making the associations where summation over k is suppressed, $k = 1, 2, 3$,

$$D_{kk} = \lambda^k, \qquad (2.37)$$

and

$$A_{ij} = u_i^j \quad \text{for} \quad i, j = 1, 2, 3. \qquad (2.38)$$

Here, the symbol **D** was chosen to represent the *diagonal* matrix. In the special case where two eigenvalues are coincident, it is possible to find two orthogonal eigenvectors. Similarly, if all three eigenvalues are coincident, it is possible to find three orthogonal eigenvectors. A proof of this result can be found in any standard text on linear algebra.

It is now useful to write

$$\mathbf{D} = \mathbf{A}^T \cdot \mathbf{\Sigma} \cdot \mathbf{A}. \tag{2.39}$$

From our previous results on similarity transformation, it follows that **D** is a representation of the stress tensor in a special coordinate system, i.e. one whose axes coincide with the eigenvectors of the original stress tensor. In this representation, we have eliminated the shear stress components and only the normal stress components survive. In continuum mechanics, this special representation for the diagonal matrix is sometimes denoted by $\mathbf{\Sigma}^*$ or by σ_{ij}^*, where the * no longer indicates complex conjugation, only that the matrix is diagonal. Since σ_{ij}^* is diagonal, its diagonal components (also called principal stresses) are sometimes denoted $\sigma_{(1)}$, $\sigma_{(2)}$, and $\sigma_{(3)}$ or by σ_I, σ_{II}, and σ_{III}. We shall employ the notationally simplest description, namely σ_1, σ_2, and σ_3. (In order to avoid unnecessary confusion, we shall not use the σ_{ij}^* notation.) For convenience, the three eigenvalues are often re-ordered so that $\sigma_1 \geq \sigma_2 \geq \sigma_3$.

Recall that the eigenvalues of a matrix remain invariant under a coordinate transformation; in the case of the stress tensor, they reflect the amount of stress exerted along each of the principal directions. Thus, any simple function of the eigenvalues is also an invariant, for example the trace of the stress tensor and its second and third powers—these are respectively $\sigma_1 + \sigma_2 + \sigma_3$, $\sigma_1^2 + \sigma_2^2 + \sigma_3^2$, $\sigma_1^3 + \sigma_2^3 + \sigma_3^3$. These latter three quantities could be employed in lieu of σ_1, σ_2, and σ_3. Similarly, the quantities $\sigma_1 + \sigma_2 + \sigma_3$, $\sigma_1\sigma_2 + \sigma_2\sigma_3 + \sigma_3\sigma_1$, and $\sigma_1\sigma_2\sigma_3$ could be regarded as invariants. These latter quantities are derived from the trace of various powers of Σ; they are denoted as $\mathbf{I}_\sigma$, $\mathbf{II}_\sigma$, and $\mathbf{III}_\sigma$. They are defined in turn by

$$\mathbf{I}_\sigma = \operatorname{tr} \mathbf{\Sigma} \tag{2.40}$$

$$\mathbf{II}_\sigma = \frac{1}{2}\left[(\operatorname{tr} \mathbf{\Sigma})^2 - \operatorname{tr}\left(\mathbf{\Sigma}^2\right)\right] \tag{2.41}$$

$$\mathbf{III}_\sigma = \det \mathbf{\Sigma}, \tag{2.42}$$

and are employed as the invariants since these terms are numerically equal to the coefficients which appear in the characteristic equation, namely

$$\sigma^3 - \mathbf{I}_\sigma \sigma^2 + \mathbf{II}_\sigma \sigma - \mathbf{III}_\sigma = 0. \tag{2.43}$$

We now proceed to derive a simple method for solving cubic equations with strictly real roots.

2.6 Solving the cubic eigenvalue equation problem

We have now shown that the eigenvalue problem associated with the stress tensor corresponds to finding the roots of a cubic polynomial. The issue of cubic polynomials has a history that spans much of antiquity ranging from the Greeks, Chinese, Indians, and Persians. However, the name Gerolamo Cardano (Nickalls, 1993; McKelvey, 1984) and his *Ars Magna* of 1545 are often associated with analytic methods for solving this problem. We present below a simplified version for finding all three roots of a cubic equation whose solutions are known *a priori* to be real-valued. Many continuum mechanics textbooks overlook this issue, which is critical to understanding three dimensional behavior. Often, students simply employ computational packages such as Matlab, Mathematica, or Maple to obtain the eigenvalues, but there is significant merit in understanding the connection between the process and the invariants as well as the Newton identities described earlier.

First, it is useful to consider the solution to a cubic equation (with $a \neq 0$) with arbitrary coefficients, namely

$$a z^3 + b z^2 + c z + d = 0. \tag{2.44}$$

For reasons that will become clear, we wish to eliminate the quadratic term from this equation; we do so by replacing z by $z' + \alpha$ where $\alpha = -b/3a$. Given our discussion of the invariants earlier, we recognize that this shift is equivalent to moving the coordinate origin for this problem to the mean value of the eigenvalues. We obtain a new equation

$$a z'^3 + c' z' + d' = 0, \tag{2.45}$$

where $c' = c + 2\alpha b + 3\alpha^2 a$ and $d' = d + c\alpha + b\alpha^2 + a\alpha^3$. We will drop the primes in (2.45) for notational convenience. Now, recall the "triple-angle" identity, which can be derived by writing $\cos \theta$ in exponential form,

$$\cos^3 \theta = \frac{1}{4} \cos 3\theta + \frac{3}{4} \cos \theta. \tag{2.46}$$

We replace z in (2.45) by $\gamma \cos \theta$, whereupon we can write

$$\gamma \cos \theta \left[\frac{3 a \gamma^2}{4} + c \right] + \left[\frac{a \gamma^3 \cos 3\theta}{4} + d \right] = 0. \tag{2.47}$$

Finally, to solve this equation, we select γ so that the first bracketed term disappears, and we then select θ so that the second bracketed term disappears. Note, however, that the solution for θ can also be changed by $\pm 2\pi/3$ giving rise to three solutions. It is important to note that in the general polynomial case the solutions for γ and for θ can be complex, but that is not in issue for us[1] since the eigenvalues of a stress tensor and, as we shall see in the next chapter, the strain tensor are real-valued. Nevertheless, what is quite remarkable here is that we can calculate *all* three solutions for a cubic polynomial analytically. In order to elucidate this process, we will consider below two simple examples. Thus, solving real-world problems in the geosciences is not as formidable as it might have seemed.

(1) Consider the polynomial

$$f(z) = (z-1)(z-2)(z-3) = z^3 - 6z^2 + 11z - 6. \quad (2.48)$$

Following the process described above, we shift the origin by $6/3 = 2$ units and employ

$$z' = z + 2, \quad (2.49)$$

whereby we obtain the reduced equation

$$z'^3 - z' = 0. \quad (2.50)$$

While we can solve this by inspection, obtaining roots at 0, and ± 1, we continue making the substitution

$$z' = \gamma \cos \theta \quad (2.51)$$

and obtain

$$0 = \frac{1}{4} \gamma^3 \cos(3\theta) + \gamma \left(\frac{3}{4} \gamma^2 - 1 \right) \cos(\theta) \quad (2.52)$$

after using the triple-angle identity. We now select $\gamma = \sqrt{4/3}$ to make the second term vanish and observe that 3θ must be $\pi/2$ of $\pi/2 + 2\pi/3$ or $\pi/2 - 2\pi/3$. Substituting these expressions back into our earlier expressions, we recover the three roots 1, 2, and 3.

(2) Consider a less transparent example

$$f(z) = z^3 - 6.2z^2 + 12.67z - 8.55. \quad (2.53)$$

As before, we proceed to obtain

$$z' = z - \frac{6.2}{3} = z - 2.0\overline{6} \quad (2.54)$$

---

[1] It is easy to see that any cubic polynomial with real coefficients must have at least one real root. The proof of this follows from plotting the behavior of a cubic polynomial from very large and negative values of z to very large and positive values of z. Since the polynomial varies continuously, it must cross the axis at least once. Further, if a root is complex, it follows that its complex conjugate is also a root. Thus, if only one root is real, then the remaining two roots form a complex conjugate pair. In that case, the exponential form or $\cosh \theta$ should be used in the calculations.

and obtain

$$0 = z'^3 - 0.14\overline{3}z' - 0.01\overline{925}. \tag{2.55}$$

Making the trigonometric substitution $z' = \gamma \cos(\theta)$, we obtain

$$0 = \gamma \left(0.75\gamma^2 - 0.14\overline{3}\right) \cos(\theta) + 0.25\gamma^3 \cos(3\theta) - 0.1\overline{925}. \tag{2.56}$$

Setting $\gamma = \sqrt{0.14\overline{3}/0.75}$, the first term is eliminated and solving for θ we obtain $\theta = 0.4371\,625\ldots$ for one root, with the other roots emerging from shifting its value by $\pm 2\pi/3 \approx \pm 2.094\,395\ldots$. Following a bit more arithmetic, the roots 1.8, 1.9, and 2.5 emerge.

We now return to questions emergent from the geometry of the stress tensor. We wish to identify the directions in which the traction is maximized and minimized.

2.7 Maximum and minimum stress values

Recall that the traction or stress vector satisfied

$$t_i^{(\hat{\mathbf{n}})} = \sigma_{ij}\,\hat{n}_j, \tag{2.57}$$

where $\hat{\mathbf{n}}$ denotes the normal vector. It is useful to decompose the stress vector into its normal and shear components (employing subscripts N and S, respectively); thus,

$$\sigma_N = \sigma_{ij}\,\hat{n}_j\,\hat{n}_i, \tag{2.58}$$

or

$$\sigma_N = \mathbf{t}^{(\hat{\mathbf{n}})} \cdot \hat{\mathbf{n}}. \tag{2.59}$$

Geometrically, it follows that

$$\sigma_N^2 + \sigma_S^2 = t_i^{(\hat{\mathbf{n}})}\,t_i^{(\hat{\mathbf{n}})}, \tag{2.60}$$

from which we can determine σ_S. A natural question that emerges is what are the *extremal* or maximum/minimum values that we can obtain for the normal stress. It follows that we are seeking the maximum value of σ_N as defined in (2.58) subject to the constraint $\hat{n}_i\hat{n}_i = 1$. This becomes a classic application of the *Lagrange multiplier method*.

We construct the function

$$f(\hat{\mathbf{n}}) = \sigma_{ij}\,\hat{n}_i\,\hat{n}_j - \sigma\left(\hat{n}_i\,\hat{n}_i - 1\right). \tag{2.61}$$

We require that the derivatives of f with respect to the $\hat{n}_i$ and with respect to σ all vanish. After some straightforward algebra and exploiting the symmetry in σ_{ij}, it follows that

$$\left(\sigma_{kj} - \sigma\, \delta_{kj}\right) \hat{n}_j = 0, \qquad (2.62)$$

which is identical to the eigenvalue problem which we have just solved. Thus, we observe that the Lagrange multiplier also assumes the role of the principal stress.

We now continue our calculation in the coordinate system corresponding with the principal axes, employing Arabic numerals as subscripts. (It is not uncommon to see Roman numerals employed in the literature in this calculation. It will be understood that our eigenvectors are normalized to unity, and we will dispense with using a circumflex " ˆ " for brevity.) Assuming that the principal stresses are ordered (as before), i.e. $\sigma_1 > \sigma_2 > \sigma_3$, we observe that

$$\sigma_N = \sigma_1 \hat{n}_1^2 + \sigma_2 \hat{n}_2^2 + \sigma_3 \hat{n}_3^2. \qquad (2.63)$$

It also follows that

$$\mathbf{t}^{(\hat{\mathbf{n}})} = \mathbf{\Sigma} \cdot \hat{\mathbf{n}} = \sigma_1 \hat{n}_1 \hat{\mathbf{e}}_1^* + \sigma_2 \hat{n}_2 \hat{\mathbf{e}}_2^* + \sigma_3 \hat{n}_3 \hat{\mathbf{e}}_3^*, \qquad (2.64)$$

where $\hat{\mathbf{e}}_i^*$ denotes the ith principal direction. Hence, the shear stress satisfies

$$\sigma_S^2 = \sigma_1^2 \hat{n}_1^2 + \sigma_2^2 \hat{n}_2^2 + \sigma_3^2 \hat{n}_3^2 - \left(\sigma_1 \hat{n}_1^2 + \sigma_2 \hat{n}_2^2 + \sigma_3 \hat{n}_3^2\right)^2. \qquad (2.65)$$

(Here, the first set of terms is just $\mathbf{t}^{(\hat{\mathbf{n}})} \cdot \mathbf{t}^{(\hat{\mathbf{n}})}$.) If we make use of $\hat{n}_i \hat{n}_i = 1$, then we can eliminate $\hat{n}_3$ leaving σ_S as a function of $\hat{n}_1$ and $\hat{n}_2$ alone.

The derivation of the solution to this problem is not especially simple, so we will outline here how to accomplish this. For convenience, we will replace $\hat{n}_1^2$ by u_1, $\hat{n}_2^2$ by u_2, and $\hat{n}_3^2$ by u_3 and note that $0 \leq u_i \leq 1$, for $i = 1, \ldots, 3$, in our expression (2.65) relating to the shear stress. We observe that this expression is now a quadratic in the u_i. A familiar result from calculus is that any continuous, differentiable function in some domain whose derivative does not vanish inside that domain must have its extrema on the boundary. In our case, the boundaries correspond to u_i being 0 or 1 for $i = 1, 2,$ and 3. In words, we are dealing with a geometrical cube. If any one of the $u_i = 1$, then the other two are identically zero and (2.65) will vanish thereby rendering the shear stress a minimum. To find the maximal shear stress, we must consider instead the case where one of the $u_i = 0$. Without loss of generality, suppose that corresponds to $i = 3$ and $u_3 = 0$. Then, we have that $u_1 + u_2 = 1$ and (2.65) can now be expressed as a function of u_1 alone and its maximum value is observed to be $\left|\frac{1}{2}(\sigma_1 - \sigma_2)\right|$ with $u_1 = u_2 = \frac{1}{2}$. We can, in like manner, now consider the remaining two cases.

Two classes of solution for the problem emerge. In the first case, the shear stress minimum, we observed that $\hat{n}_1 = \hat{n}_2 = 0$ and $\hat{n}_3 = \pm 1$. This is expected since this defines a principal plane, thereby leaving a vanishing shear stress. We also obtain a vanishing shear stress when either $u_1 = 1$ or $u_2 = 2$. To obtain the maximum shear stress, we extend the algebra described above and the solution can be represented in tabular form:

$$\begin{aligned} \hat{n}_1 &= 0 & \hat{n}_2 &= \pm\frac{1}{\sqrt{2}} & \hat{n}_3 &= \pm\frac{1}{\sqrt{2}} & \sigma_S &= \left|\frac{1}{2}(\sigma_2 - \sigma_3)\right| \\ \hat{n}_1 &= \pm\frac{1}{\sqrt{2}} & \hat{n}_2 &= 0 & \hat{n}_3 &= \pm\frac{1}{\sqrt{2}} & \sigma_S &= \left|\frac{1}{2}(\sigma_3 - \sigma_1)\right|. \\ \hat{n}_1 &= \pm\frac{1}{\sqrt{2}} & \hat{n}_2 &= \pm\frac{1}{\sqrt{2}} & \hat{n}_3 &= 0 & \sigma_S &= \left|\frac{1}{2}(\sigma_1 - \sigma_2)\right| \end{aligned} \quad (2.66)$$

Owing to the ordering of the principal stresses, it follows that the largest shear stress value is

$$\sigma_{S(max)} = \left|\frac{1}{2}(\sigma_3 - \sigma_1)\right|. \tag{2.67}$$

Before proceeding with further mathematical development, it is important to consider the physical meaning of these quantities, particularly in the context of geoscience applications.

A particularly important example relates to the earthquake mechanism. The Great San Francisco Earthquake of 1906 provided the evidence employed by Reid (1911) to develop his *elastic rebound theory* of earthquakes. Earthquakes are associated with large fractures or faults in the Earth's crust and upper mantle (Press and Siever, 1986). In essence, imagine that a fault separates two masses of Earth materials which are being pushed together by applied stress, the outcome of plate tectonic forces due to deep Earth motion. The traction or stress that emerges on the fault surface will have two components, as discussed above, which we can identify with the normal stress and the shear stress. An additional physical process now enters into this picture, namely *friction*. From our day-to-day experience, we expect that the presence of a normal stress will result in a resistive frictional stress that will oppose the shear stress that is present. Our ideas relating to friction (Scholz, 2002) go back to Leonardo da Vinci, Sir Isaac Newton in his *Principia*, and especially Guillaume Amontons. Amontons' two (empirical) laws occupy a fundamental role in continuum mechanics and related disciplines.

Law 1. The frictional force is independent of the size of the surfaces in contact.
Law 2. Friction is proportional to the normal load.

(Sometimes, a third law is added to these, albeit due to Coulomb: kinetic friction is independent of the sliding velocity.) Much of our understanding of friction has emerged from laboratory experiments and introduces the notion of *asperities*, essentially protrusions on the two surfaces. In these phenomenological theories, a

42   *Stress principles*

coefficient of friction, often designated μ, describes the fraction of the normal stress that can manifest as a frictional stress that opposes the shear stress. Therefore, if $\mu\sigma_N > \sigma_S$, then *slip* between the two surfaces will not occur. An elaboration of these ideas can be found in the texts by Turcotte and Schubert (2002) as well as Sleep and Fujita (1997). Recently, Müser *et al.* (2001) has developed a simple microscopic theory to explain Amontons' empirical laws.

2.8 Mohr's circles

Consider as before the normal and shear stress representations in the coordinate system defined by the principal directions. We have three conditions emerge upon the normal vector $\hat{\mathbf{n}}$, namely

$$\sigma_N = \sigma_1 \hat{n}_1^2 + \sigma_2 \hat{n}_2^2 + \sigma_3 \hat{n}_3^2 \tag{2.68}$$

$$\sigma_N^2 + \sigma_S^2 = \sigma_1^2 \hat{n}_1^2 + \sigma_2^2 \hat{n}_2^2 + \sigma_3^2 \hat{n}_3^2, \tag{2.69}$$

and the normalization condition

$$\hat{n}_1^2 + \hat{n}_2^2 + \hat{n}_3^2 = 1. \tag{2.70}$$

We can regard these three equations as defining the three direction cosines $\hat{n}_i$, for $i = 1, 2, 3$. To simultaneously satisfy these three equations, we must obtain the intersection of the three ellipses in the $\hat{n}_1$–$\hat{n}_2$–$\hat{n}_3$ coordinate system that they describe. An algebraically simpler way to proceed is to employ the u_i notation introduced earlier, rendering these three expression a linear system of equations in the u_i. In particular, we obtain

$$\hat{n}_1^2 = \frac{(\sigma_N - \sigma_2)(\sigma_N - \sigma_3) + \sigma_S^2}{(\sigma_1 - \sigma_2)(\sigma_1 - \sigma_3)} \tag{2.71}$$

$$\hat{n}_2^2 = \frac{(\sigma_N - \sigma_3)(\sigma_N - \sigma_1) + \sigma_S^2}{(\sigma_2 - \sigma_3)(\sigma_2 - \sigma_1)} \tag{2.72}$$

$$\hat{n}_3^2 = \frac{(\sigma_N - \sigma_1)(\sigma_N - \sigma_2) + \sigma_S^2}{(\sigma_3 - \sigma_1)(\sigma_3 - \sigma_2)}. \tag{2.73}$$

In this system, σ_1, σ_2, and σ_3 are known, but σ_N and σ_S may be regarded as functions of the direction cosines $\hat{n}_i$. The first and third of these equations have positive denominators, making it necessary that their respective numerators also be positive. On the other hand, the second of these equations has a negative denominator making it necessary that its numerator also be negative. Meanwhile, the left-hand sides of each of the three equations are necessarily positive or zero. Each of these bounds gives rise to the existence of circles which bound the solution in the σ_N–σ_S

2.8 Mohr's circles

plane which bounds the admissible solutions. Hence, we derive the three conditions which define these circles:

$$\left[\sigma_N - \frac{1}{2}(\sigma_2 + \sigma_3)\right]^2 + \sigma_S^2 \geq \left[\frac{1}{2}(\sigma_2 - \sigma_3)\right]^2 \tag{2.74}$$

$$\left[\sigma_N - \frac{1}{2}(\sigma_1 + \sigma_3)\right]^2 + \sigma_S^2 \leq \left[\frac{1}{2}(\sigma_1 - \sigma_3)\right]^2 \tag{2.75}$$

$$\left[\sigma_N - \frac{1}{2}(\sigma_1 + \sigma_2)\right]^2 + \sigma_S^2 \geq \left[\frac{1}{2}(\sigma_1 - \sigma_2)\right]^2. \tag{2.76}$$

We show these circles in the diagram below (figure 2.4), where

$$C_I = \left|\frac{1}{2}(\sigma_2 - \sigma_3)\right| \tag{2.77}$$

$$C_{II} = \left|\frac{1}{2}(\sigma_3 - \sigma_1)\right| \tag{2.78}$$

$$C_{III} = \left|\frac{1}{2}(\sigma_1 - \sigma_2)\right|. \tag{2.79}$$

The three circles shown are known as *Mohr's circles for stress* and the shaded region describes all admissible pairs of σ_N and σ_S. The three circles are identified by subscripts which correspond to the three equations above.

Suppose we have an admissible pair σ_N and σ_S and wish to find the corresponding set of $\hat{n}_1$, $\hat{n}_2$, and $\hat{n}_3$. This we do according to (2.71), (2.72), and (2.73). We eliminate the 8-fold indeterminacy in the solutions by selecting each of the direction cosines $\hat{n}_1$, $\hat{n}_2$, and $\hat{n}_3$ to be positive, that is we select the direction cosines to

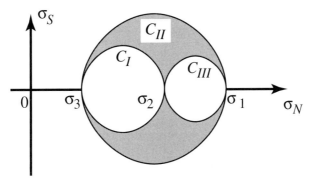

Figure 2.4 Geometry of Mohr circles.

be in the *first octant* of the $\hat{n}_1$–$\hat{n}_2$–$\hat{n}_3$ space. Hence, an equivalent representation emerges if we select angles $0 \leq \phi_i \leq \pi/2$, for $i = 1, 2, 3$ according to

$$\phi_i = \cos^{-1}(\hat{n}_i). \tag{2.80}$$

This representation facilitates other geometrical manipulations. For example, if we select ϕ_3, and allow ϕ_1 and ϕ_2 to vary, equation (2.64) can be rewritten

$$(\sigma_N - \sigma_1)(\sigma_N - \sigma_2) + \sigma_S^2 = \cos^2 \phi_3 (\sigma_3 - \sigma_1) \cdot (\sigma_3 - \sigma_2), \tag{2.81}$$

and then in the form of the equation for a circle

$$\left[\sigma_N - \frac{1}{2}(\sigma_1 + \sigma_2)\right]^2 + \sigma_S^2 = \frac{1}{4}(\sigma_1 - \sigma_2)^2 + \cos^2 \phi_3 \cdot (\sigma_3 - \sigma_1)(\sigma_3 - \sigma_2). \tag{2.82}$$

This circle has the same origin as that shown in the figure with radius C_{III}, but whose radius lies outside the one shown (unless $\phi_3 = \pi/2$) – however, admissible solutions for σ_N and σ_S emerge outside the circle plotted but within the shaded region. Similar manipulations can be performed based upon fixing the values of ϕ_1 or ϕ_2. Finally, it should be emphasized that once the direction cosines have been identified, we can calculate the stress vector directly from the product of the stress tensor with the associated normal vector.

2.9 Plane, deviator, spherical, and octahedral stress

The case where one principal stress value vanishes – known as a state of *plane stress* – deserves mention. In that case which emerges in two-dimensional problems, we identify the plane specified by the two non-zero principal stresses as the *designated plane*. Since one of σ_1, σ_2, or σ_3 vanish, it then follows that the σ_S axis comes in contact with two of the circles. We illustrate in figure 2.5 below the case where $\sigma_3 = 0$.

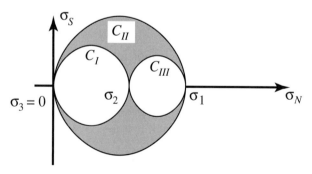

Figure 2.5 Mohr circle and plane stress.

2.9 Plane, deviator, spherical, and octahedral stress

Suppose that the corresponding principal direction is aligned along the $\hat{\mathbf{e}}_3$ axis. Then the stress tensor has the form

$$\sigma = \begin{pmatrix} \sigma_{11} & \sigma_{12} & 0 \\ \sigma_{21} & \sigma_{22} & 0 \\ 0 & 0 & 0 \end{pmatrix} \quad (2.83)$$

and the corresponding Mohr circle has the form

$$\left(\sigma_N - \frac{\sigma_{11} + \sigma_{22}}{2}\right)^2 + \sigma_S^2 = \left(\frac{\sigma_{11} - \sigma_{22}}{2}\right)^2 + \sigma_{12}^2. \quad (2.84)$$

This equation establishes that the Mohr circle has its center at $\sigma_N = \frac{1}{2}(\sigma_{11} + \sigma_{22})$ and $\sigma_S = 0$. In addition, the simpler form of the stress tensor allows us to express the other principal stress values, namely σ_1 and σ_2, as $\frac{1}{2}(\sigma_{11} + \sigma_{22}) \pm \sqrt{\left[\frac{1}{2}(\sigma_{11} - \sigma_{22})\right]^2 + \sigma_{12}^2}$. An additional result that emerges from this figure, either by inspection or a simple calculation, is that the ratio of the (absolute value of the) shear stress σ_S to the normal stress σ_N is maximized when they are equal. This occurs when $|\hat{n}_1| = |\hat{n}_2| = \sqrt{2}/2$. In other words, the greatest likelihood of slip in this kind of situation occurs when the compressive force is approximately 45° from the fault surface. This helps explain, for example, why cracks in building materials in the aftermath of earthquakes often appear at an angle of 45° from the vertical.

In the nomenclature, three other descriptions of the state of stress appear. The first pair of these relates to the arithmetic mean of the normal stresses,

$$\sigma_M = \frac{1}{3}(\sigma_{11} + \sigma_{22} + \sigma_{33}) = \frac{1}{3}\sigma_{ii}, \quad (2.85)$$

which is called the mean normal stress. Recalling that the trace of a matrix is invariant, we note that the mean normal stress is also the mean of the principal stresses.

Thus, in the case where the three principal stresses are equal (and hence equal σ_M), we have a *spherical state of stress* represented by the diagonal matrix

$$[\sigma_{ij}] = \begin{pmatrix} \sigma_M & 0 & 0 \\ 0 & \sigma_M & 0 \\ 0 & 0 & \sigma_M \end{pmatrix}. \quad (2.86)$$

Since the eigenvalues of this matrix are completely degenerate, we may arbitrarily select any directions for the principal directions. A spherical state of stress applies in problems which are microscopically isotropic, such as in a fluid in a rest or *hydrostatic* state. Here, the static pressure satisfies $p_0 = -\sigma_M$.

Every state of stress σ_{ij} may be decomposed into a spherical portion and a portion S_{ij} known as the *deviator stress* according to

$$\sigma_{ij} = S_{ij} + \delta_{ij}\sigma_M = S_{ij} + \frac{1}{3}\delta_{ij}\sigma_{kk}. \quad (2.87)$$

It follows that the trace of the deviator stress vanishes so that the characteristic equation now has the form $S^3 + II_S S - III_S = 0$, and that the eigenvalues of $\mathbf{S}$ are just $\sigma_1 - \sigma_M$, $\sigma_2 - \sigma_M$, and $\sigma_3 - \sigma_M$. The principal directions of the deviator stress are simply those of the original stress tensor.

Finally, there are some applications where it is convenient to select a unit normal which makes equal angles with the principal stress directions. Assuming that we select the coordinate axes to coincide with the principal directions, the stress vector then has the form

$$\mathbf{t}^{(\hat{\mathbf{n}})} = \sigma \cdot \hat{\mathbf{n}} = \frac{\sigma_1 \hat{\mathbf{e}}_1 + \sigma_2 \hat{\mathbf{e}}_2 + \sigma_3 \hat{\mathbf{e}}_3}{\sqrt{3}}, \tag{2.88}$$

whose normal stress (i.e. its component in the direction of $\hat{\mathbf{n}}$) is just

$$\sigma_N = \mathbf{t}^{(\hat{\mathbf{n}})} \cdot \hat{\mathbf{n}} = \frac{1}{3}[\sigma_1 + \sigma_2 + \sigma_3] = \frac{1}{3}\sigma_{ii}. \tag{2.89}$$

In this case, the (octahedral) shear stress is

$$\begin{aligned} \sigma_{oct}^2 &= \mathbf{t}^{(\hat{\mathbf{n}})} \cdot \mathbf{t}^{(\hat{\mathbf{n}})} - \sigma_N^2 \\ &= \frac{1}{3}[\sigma_1^2 + \sigma_2^3 + \sigma_3^2] - \frac{1}{9}[\sigma_1 + \sigma_2 + \sigma_3]^2. \end{aligned} \tag{2.90}$$

This can also be expressed in the form

$$\begin{aligned} \sigma_{oct} &= \frac{1}{3}\sqrt{[\sigma_1 - \sigma_2]^2 + [\sigma_2 - \sigma_3]^2 + [\sigma_3 - \sigma_1]^2} \\ &= \sqrt{\frac{S_1^2 + S_2^2 + S_3^2}{3}}, \end{aligned} \tag{2.91}$$

which is the root-mean-squared value of the deviatoric stresses. These terms are sometimes encountered in engineering applications. While the mathematics that we have encountered is somewhat laborious, we have shown in this chapter that the analysis of three-dimensional problems does not present any fundamental challenges. A helpful resource in addressing these practical mathematical questions can be found in Oertel (1996). Moreover, we have seen that the outcome of the mathematics has an immediate geometrical and physically intuitive appeal. We now proceed to a simple description of kinematics.

Exercises

2.1 Consider the stress tensor

$$[\sigma_{ij}] = \begin{pmatrix} 2 & 2 & 4 \\ 2 & 5 & 8 \\ 4 & 8 & 17 \end{pmatrix}$$

evaluated at the coordinate origin P. What is the associated characteristic equation and its eigenvalues?

2.2 Consider the stress tensor

$$[\sigma_{ij}] = \begin{pmatrix} 0 & 1 & 1 \\ 1 & 0 & 1 \\ 1 & 1 & 0 \end{pmatrix}$$

evaluated at the coordinate origin P. What is the associated characteristic equation and its eigenvalues?

2.3 Consider the stress tensor

$$[\sigma_{ij}] = \begin{pmatrix} 4 & 1 & -1 \\ 1 & 3 & 0 \\ -1 & 0 & 5 \end{pmatrix}$$

evaluated at the coordinate origin P. What is the associated characteristic equation and eigenvalues?

2.4 Consider the stress tensor

$$[\sigma_{ij}] = \begin{pmatrix} 2 & 1 & -1 \\ 1 & 3 & 0 \\ -1 & 0 & 3 \end{pmatrix}$$

evaluated at the coordinate origin P. What is the associated characteristic equation and eigenvalues? Notice how differently this matrix behaves in contrast with the previous problem. If you performed this calculation as described in the text, everything should work out nicely. However, if you performed it in Mathematica or Maple, your result may look very unusual. Are these eigenvalues genuinely complex?

2.5 Determine the stress vector on the plane at P parallel to plane BGE and $BGFC$ in figure 2.6. Determine the normal and the shear stress components on the plane $BGFC$ in this figure.

2.6 Decompose the stress tensor

$$[\sigma_{ij}] = \begin{pmatrix} 6 & -10 & 0 \\ -10 & 3 & 30 \\ 0 & 30 & -24 \end{pmatrix}$$

into its spherical and deviator parts and determine the principal deviator stresses.

2.7 Show that the normal component of the stress vector on the octahedral plane is equal to one third the first invariant (i.e. the trace) of the stress tensor.

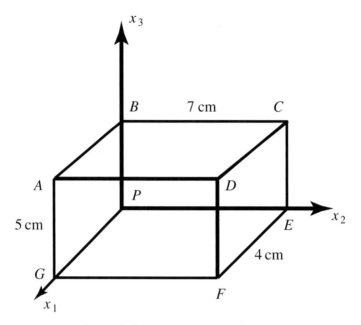

Figure 2.6 Geometry for exercise 2.5.

2.8 Sketch the Mohr's circles and determine the maximum shear stress for each of the following stress states

$$\text{(a)} \quad [\sigma_{ij}] = \begin{pmatrix} \tau & \tau & 0 \\ \tau & \tau & 0 \\ 0 & 0 & 0 \end{pmatrix}$$

and

$$\text{(b)} \quad [\sigma_{ij}] = \begin{pmatrix} \tau & 0 & 0 \\ 0 & -\tau & 0 \\ 0 & 0 & -2\tau \end{pmatrix}.$$

2.9 In a particular continuum, the stress field is given by the tensor field

$$[\sigma_{ij}] = \begin{pmatrix} x_1^2 x_2 & (1 - x_2^2) x_1 & 0 \\ (1 - x_2^2) x_1 & (x_2^3 - 3 x_2)/3 & 0 \\ 0 & 0 & 2 x_3^2 \end{pmatrix}.$$

Determine (a) the body force distribution if the equilibrium equations are to be satisfied throughout the field; (b) the principal stress values at the point $P(a, 0, \sqrt{2}a)$; (c) the maximum shear stress at P; (d) the principal deviatoric stress at P.

3

Deformation and motion

3.1 Coordinates and deformation

An essential problem in continuum mechanics is the description of the evolution of a material body, whether composed of a solid, liquid, or gas. This may be facilitated by attaching a coordinate **X**, which serves as a label, identifying every point in that material. These points can correspond to real particles or, alternatively, describe idealized particles. Further, we assume that the position of that particle can be described at any instant by a mapping Θ according to

$$\mathbf{x} = \Theta(\mathbf{X}), \qquad (3.1)$$

where Θ is a vector function which assigns the position **x** relative to some origin of each particle identified by **X** in the body. We assume that this mapping is invertible, i.e. the mapping

$$\mathbf{X} = \Theta^{-1}(\mathbf{x}) \qquad (3.2)$$

exists and that Θ is differentiable almost everywhere. While our prescription here is relatively mathematical, it conforms with essentially all physical applications that we might wish to pursue. We have chosen to employ the modern concept of a mapping in approaching issues of deformation; this methodology is very similar to that employed by authors such as Chadwick (1999), Fung (1965), Gurtin (1981), Mase and Mase (1990), and Narasimhan (1993).

The change in configuration is the result of the *displacement* of the body. For example, *rigid body displacement* consists of a simultaneous translation and rotation which produces a new configuration with no change in the size or shape of the body. This is in marked contrast with a *deformation* which will produce both.

Helmholtz (1868) established that the most general motion of a sufficiently small element of a deformable body can be represented as the sum of a translation, a rotation, and an extension (or contraction) in three mutually orthogonal directions. It is useful to review this proof here as it is of fundamental importance to the notion of strain. We begin by denoting by $\Theta^{(0)}$ the displaced position of a particle originally at the origin $\mathbf{0} = \mathbf{X}$, namely $\Theta^{(0)} = \Theta(\mathbf{0})$. Then, introducing a Taylor series, it follows that

$$x_i = \Theta_i^{(0)} + \frac{\partial \Theta_i}{\partial X_j} X_j + O(X^2), \tag{3.3}$$

where the partial derivatives are evaluated at $\mathbf{X} = \mathbf{0}$. We will rewrite this expression in the following form:

$$x_i = \Theta_i^{(0)} + \left[\frac{1}{2}\frac{\partial \Theta_i}{\partial X_j} - \frac{1}{2}\frac{\partial \Theta_j}{\partial X_i}\right] X_j + \left[\frac{1}{2}\frac{\partial \Theta_i}{\partial X_j} + \frac{1}{2}\frac{\partial \Theta_j}{\partial X_i}\right] X_j + O(X^2) \tag{3.4}$$

and each of these terms has a simple physical interpretation.

The first of these terms we recognize as a translation. The second of these terms has the form $\epsilon_{ijk} v_j X_k$ where we define v_j by

$$v_j = -\frac{1}{2} \epsilon_{jmn} \frac{\partial \Theta_m}{\partial X_n}, \tag{3.5}$$

which we recognize as a rotation, i.e. by virtue of being orthogonal to $\mathbf{X}$. Finally, the third term introduces a tensor product, where

$$\chi_{ij} = \frac{1}{2}\left[\frac{\partial \Theta_i}{\partial X_j} + \frac{\partial \Theta_j}{\partial X_i}\right] \tag{3.6}$$

is a real-valued symmetric tensor which we will relate soon to the *strain tensor*. The symmetry property of the strain tensor we will soon see is of fundamental importance. From what we have already learned, we know that it can be diagonalized and thereby represents an extension or contraction in three principal directions. Thus, we can now write (3.3) as

$$x_i = \Theta_i^{(0)} + \epsilon_{ijk} v_j X_k + \chi_{ij} X_j + O(X^2). \tag{3.7}$$

The first and second terms, describing respectively a displacement and a rotation, are often ignored since they do not describe a change in the material properties and can be accommodated by a simple coordinate transformation.

Although the third term is generally employed to describe the effect of deformation, it is important to note that it introduces approximations that are of the same order in the expansion. This can be appreciated by looking at the change in the length of the displacement of a point $\mathbf{x}$ from its original position Θ – we define the *displacement vector*

$$u_i \equiv x_i - X_i = \Theta_i(\mathbf{X}) - X_i. \tag{3.8}$$

Since we are now referring the variables to material coordinates, we are employing a *material* or *Lagrangian* description (Segel and Handelman, 1987). Consider two

3.1 Coordinates and deformation

initially nearby points, whose separation vector is denoted by its components dX_i. After the deformation, their separation distance then becomes $dx_i = dX_i + du_i$. The original distance between the points was $dX = \sqrt{dX_1^2 + dX_2^2 + dX_3^2}$ and the new distance after the deformation is $dx = \sqrt{dx_1^2 + dx_2^2 + dx_3^2}$. (By refraining from the use of curvilinear coordinates, we have avoided the need to introduce a *metric* for the coordinate system and the complications that produces.) Substituting $du_i = (\partial u_i/\partial X_k)\, dX_k$ and using $dx_i = dX_i + du_i$, we can write

$$dx^2 = dX^2 + 2\frac{\partial u_i}{\partial X_k} dX_i\, dX_k + \frac{\partial u_i}{\partial X_k}\frac{\partial u_i}{\partial X_l} dX_k\, dX_l. \qquad (3.9)$$

Since we sum over the second term in both the i and k indices, we can replace it by $\frac{\partial u_i}{\partial X_k} dX_i\, dX_k + \frac{\partial u_k}{\partial X_i} dX_i\, dX_k$ permitting us to write

$$dx^2 = dX^2 + 2\varepsilon_{ik}\, dX_i\, dX_k, \qquad (3.10)$$

where the tensor ε_{ik} was defined by

$$\varepsilon_{ik} = \frac{1}{2}\left(\frac{\partial u_i}{\partial X_k} + \frac{\partial u_k}{\partial X_i} + \frac{\partial u_l}{\partial X_i}\frac{\partial u_l}{\partial X_k}\right). \qquad (3.11)$$

Note that $\varepsilon_{ij} = \chi_{ij} - \delta_{ij}$ is a consequence of $u_i(\mathbf{X}) = \Theta_i(\mathbf{X}) - X_i$. The quantity ε_{ij} is the *strain tensor*, although most descriptions ignore the nonlinear term. This is a valid approximation where the angles and relative extensions of local deformation are small. There are, however, situations with pulse-like propagation features where the nonlinear term cannot be ignored.

The motion of a body is a continuous time sequence of displacements that carries the particles $\mathbf{X}$ into various configurations in a stationary spatial coordinate system. This can be expressed by the modified equation

$$\mathbf{x} = \Theta(\mathbf{X}, t), \qquad (3.12)$$

which gives the position $\mathbf{x}$ for each particle $\mathbf{X}$ for all times t. We distinguish between the *reference*, i.e. original at $t = 0$, and *current* configurations of a body in this way. This distinction is helpful in developing the notion of strain as it reflects the change from the *undeformed configuration* into the *deformed configuration*. Although this designation is not natural for fluids, it is useful in that it permits us to define a *velocity field* for a flow. In particular, we will see that the quantity $\partial \Theta/\partial t$ provides a natural mechanism to describe the evolution of a fluid. These geometrical concepts provide an essential tool for following the kinematic evolution of Earth materials, independent of whether we are describing the deformation of rocks or the flow of geophysical fluids.

Given that **X** is both a label and defines the original position of a particle, it is possible to define the evolution of a particle's position by (3.12) or, using the form most often employed in continuum mechanics,

$$x_i = x_i(X, t) \quad \text{or} \quad \mathbf{x} = \mathbf{x}(\mathbf{X}, t). \tag{3.13}$$

Here, **x** means either the position of the particle at this time *or* a function of its original position and time. In order to preserve the convenient notion that **X** defines the original position of the particle, we employ

$$\mathbf{x} = \mathbf{x}(\mathbf{X}, 0) = \mathbf{X}, \tag{3.14}$$

at time $t = 0$. Similarly, we denote by

$$X_i = X_i(x, t) \quad \text{or} \quad \mathbf{X} = \mathbf{X}(\mathbf{x}, t), \tag{3.15}$$

the inverse mapping taking the current position of a particle into its original position or its label. Finally, it is important to note that these mappings exist only if the Jacobian J defined by

$$J = \det \begin{pmatrix} \dfrac{\partial x_1}{\partial X_1} & \dfrac{\partial x_1}{\partial X_2} & \dfrac{\partial x_1}{\partial X_3} \\ \dfrac{\partial x_2}{\partial X_1} & \dfrac{\partial x_2}{\partial X_2} & \dfrac{\partial x_2}{\partial X_3} \\ \dfrac{\partial x_3}{\partial X_1} & \dfrac{\partial x_3}{\partial X_2} & \dfrac{\partial x_3}{\partial X_3} \end{pmatrix} \tag{3.16}$$

of the transformation $\mathbf{x}(\mathbf{X}, t)$ is non-vanishing. Physically, this can never happen, but it is an issue germane to computer simulations. For example, a vanishing Jacobian identifies that some shock-like behavior is possibly present. Pursuing this idea for a specific particle **X**, it follows that $\mathbf{x} = \mathbf{x}(\mathbf{X}, t)$ defines the path or trajectory of that particle as a function of time t. The velocity, then, satisfies

$$\mathbf{v} = \frac{d\mathbf{x}}{dt} = \left[\frac{\partial \mathbf{x}(\mathbf{X}, t)}{\partial t} \right]_\mathbf{X}. \tag{3.17}$$

Similarly, the acceleration satisfies

$$\mathbf{a} = \frac{d^2\mathbf{x}}{dt^2} = \left[\frac{\partial^2 \mathbf{x}(\mathbf{X}, t)}{\partial t^2} \right]_\mathbf{X}. \tag{3.18}$$

Finally, it is useful now to introduce the notion of *Eulerian* and *Lagrangian* descriptions of particle motion. The essential point here is that we have two equivalent descriptions available for particle motion, one based on the current position of a particle and one based on the particle's label, i.e. original position. For example, in describing the density or velocity fields of a particle, which we usually think of in terms of $\mathbf{v}(\mathbf{x}, t)$ or $\rho(\mathbf{x}, t)$, we are using the Eulerian description. On the other hand, we could have employed, instead, $\mathbf{v}(\mathbf{X}, t)$ or $\rho(\mathbf{X}, t)$, where we associate the velocity and density with the particle's label or initial position. This Lagrangian description can be invaluable in treating problems where some material property is preserved, for example, the entropy (density) at a point. In such situations, the entropy is an equivalent label and can provide immense simplifications in solving complex continuum evolution problems. Corresponding to the Eulerian and Lagrangian descriptions, derivatives with respect to coordinates may have one of two forms. The familiar case is that of taking derivatives with respect to the spatial coordinate $\mathbf{x}$, for example the density gradient with components $\partial \rho / \partial x_i$. On the other hand, we can take derivatives with respect to the Lagrangian coordinate $\mathbf{X}$ which we refer to as a *material derivative*. We do this since it possibly relates to variability with respect to some material property such as in the entropy example above. Also, when we take derivatives with respect to time, it is essential to note whether the spatial or material coordinates are held fixed.

In order to convert between the two descriptions, it is essential that the chain rule be employed, and employed with caution. One particular example comes from considering the meaning of $\partial f / \partial t$ which has different meanings in the two coordinate description – depending on the coordinates, either $\mathbf{X}$ or $\mathbf{x}$ is held fixed. Thus, suppose we wish to calculate $\partial f / \partial t$ in Lagrangian coordinates in terms of the usual spatial description. Recall that

$$df = \left.\frac{\partial f}{\partial t}\right|_{\mathbf{x}} dt + \left.\frac{\partial f}{\partial x_i}\right|_t dx_i. \qquad (3.19)$$

Using this relation we find

$$\left.\frac{\partial f}{\partial t}\right|_{\mathbf{X}} = \left.\frac{\partial f}{\partial t}\right|_{\mathbf{x}} + \left.\frac{\partial f}{\partial x_i}\right|_t \left.\frac{\partial x_i}{\partial t}\right|_{\mathbf{X}}. \qquad (3.20)$$

We recognize from (3.17) that the last term corresponds to components v_i of the velocity vector and, therefore, that the material derivative on the left-hand side is the *total* or *convective* derivative sometimes denoted simply as df/dt or Df/Dt. Thus, our equation can be written

$$\frac{df}{dt} = \frac{\partial f}{\partial t} + v_i \frac{\partial f}{\partial x_i} = \frac{\partial f}{\partial t} + \mathbf{v} \cdot \nabla_{\mathbf{x}} f. \qquad (3.21)$$

Here, we employed the $\mathbf{x}$ subscript on the gradient operator to identify that differentiation is taking place with respect to the spatial coordinate.

3.2 Strain tensor

In our previous discussion, we have already identified several tensor quantities. The first of these, namely $\partial \Theta_i / \partial X_j$, is sometimes referred to as the *deformation gradient tensor*, although it is not often employed in the literature. The quantity defined in (3.11), which we repeat here

$$\varepsilon_{ik} = \frac{1}{2}\left(\frac{\partial u_i}{\partial X_k} + \frac{\partial u_k}{\partial X_i} + \frac{\partial u_l}{\partial X_i}\frac{\partial u_l}{\partial X_k}\right), \quad (3.22)$$

is called the *Lagrangian finite strain tensor*. The term "Lagrangian" is applied since material derivatives are employed – the symbol E_{ij} is sometimes used instead of ε_{ij}. Other tensors that appear include the *Green's deformation tensor* defined by $\frac{\partial x_i}{\partial X_j}\frac{\partial x_i}{\partial X_k}$, and the *Cauchy deformation tensor* defined by $\frac{\partial X_i}{\partial x_j}\frac{\partial X_i}{\partial x_k}$. Finally, we define the *Eulerian finite strain tensor* by

$$e_{ik} = \frac{1}{2}\left(\frac{\partial u_i}{\partial x_k} + \frac{\partial u_k}{\partial x_i} - \frac{\partial u_l}{\partial x_i}\frac{\partial u_l}{\partial x_k}\right). \quad (3.23)$$

The latter emerges in defining the change in the length analogous to (3.10), namely

$$dx^2 = dX^2 + 2e_{ik}\,dx_i\,dx_k. \quad (3.24)$$

With this notation, we now proceed to develop the associated theory.

3.3 Linearized deformation theory

In situations where $\|\partial u_i/\partial X_i\| \ll 1$, it follows that we can approximate ε_{ij} by

$$\varepsilon_{ij} = \frac{1}{2}\left(\frac{\partial u_i}{\partial X_j} + \frac{\partial u_j}{\partial X_i}\right). \quad (3.25)$$

It should be noted that there are several situations where this approximation does not apply. These include (1) discontinuities in the physical medium, e.g. fractures; (2) discontinuities in the properties of the medium, e.g. shocks; and (3) deformations giving rise to large-angle shears. While the first two situations appear obvious, we will demonstrate the relevance of the latter restriction later. Noting that

$$\frac{\partial u_i}{\partial X_j} = \frac{\partial u_i}{\partial x_k}\frac{\partial x_k}{\partial X_j} = \frac{\partial u_i}{\partial x_k}\left(\frac{\partial u_k}{\partial X_j} + \delta_{jk}\right) \approx \frac{\partial u_i}{\partial x_k}\delta_{jk} = \frac{\partial u_i}{\partial x_j}, \quad (3.26)$$

we find that to the same degree of approximation

$$e_{ij} = \varepsilon_{ij}, \quad (3.27)$$

and we will cease to distinguish between the Lagrangian and Eulerian form of stress tensor in the case of infinitesimal deformations.

Recalling that ε_{ij} is symmetric, we observe that its eigenvalues are real and that it possesses three eigenvectors and a diagonal representation

$$\begin{pmatrix} \varepsilon_1 & 0 & 0 \\ 0 & \varepsilon_2 & 0 \\ 0 & 0 & \varepsilon_3 \end{pmatrix}, \tag{3.28}$$

together with strain invariants

$$I_\varepsilon = \varepsilon_{ii} = \text{tr}(\boldsymbol{\varepsilon}) = \varepsilon_1 + \varepsilon_2 + \varepsilon_3, \tag{3.29}$$

$$II_\varepsilon = \frac{1}{2}\left(\varepsilon_{ii}\,\varepsilon_{jj} - \varepsilon_{ij}\,\varepsilon_{ji}\right) = \varepsilon_1\,\varepsilon_2 + \varepsilon_2\,\varepsilon_3 + \varepsilon_1\,\varepsilon_3, \tag{3.30}$$

and

$$III_\varepsilon = \epsilon_{ijk}\,\varepsilon_{1i}\,\varepsilon_{2j}\,\varepsilon_{3k} = \varepsilon_1\,\varepsilon_2\,\varepsilon_3. \tag{3.31}$$

We now wish to apply this formalism to situations that correspond physically to "stretching" a rubber sheet or "squeezing" a piece of putty, in two or three dimensions, respectively. This visualization is important in developing an intuitive understanding of the meaning of our calculations.

In this infinitesimal deformation limit, we can obtain simple expressions for the change in length dX and dx we considered earlier. In particular, recalling

$$dx^2 - dX^2 = 2\,\varepsilon_{ij}\,dX_i\,dX_j, \tag{3.32}$$

it follows that

$$\frac{dx - dX}{dX}\frac{dx + dX}{dX} \approx 2\frac{dx - dX}{dX} = 2\,\varepsilon_{ij}\frac{dX_i}{dX}\frac{dX_j}{dX} = 2\hat{\mathbf{N}}\cdot\boldsymbol{\varepsilon}\cdot\hat{\mathbf{N}}, \tag{3.33}$$

where the components of the unit vector $\hat{\mathbf{N}}$ are defined by $\hat{N}_i = dX_i/dX$. The left-hand side of the previous equation defines the relative change in the length per unit of the original length. A case of particular interest emerges when $\hat{\mathbf{N}}$ is aligned along one of the principal directions of the strain tensor. Then, this relative change in the length is observed to be equal to one of the principal strains. Thus, we see physically that the principal strains define the relative change in length, either in extension or contraction, along the principal directions. We call this the *longitudinal strain* or the *normal strain* and denote it by $e_{(\hat{\mathbf{N}})}$. In the more general case where our coordinates are arbitrary, we see that the diagonal elements of the strain tensor represent the normal strains along each of the selected coordinate axes.

In order to understand the meaning of the off-diagonal elements, consider two initial differential vectors $d\mathbf{X}^{(1)}$ and $d\mathbf{X}^{(2)}$ and their deformed values $d\mathbf{x}^{(1)}$ and $d\mathbf{x}^{(2)}$. After a little algebra, we obtain

$$dx^{(1)} \cdot dx^{(2)} = d\mathbf{X}^{(1)} \cdot d\mathbf{X}^{(2)} + d\mathbf{X}^{(1)} \cdot 2\boldsymbol{\varepsilon} \cdot d\mathbf{X}^{(2)}. \tag{3.34}$$

Now we select that the two initial differential vectors be orthogonal to each other, whereupon (3.34) becomes

$$dx^{(1)} \cdot dx^{(2)} = d\mathbf{X}^{(1)} \cdot 2\boldsymbol{\varepsilon} \cdot d\mathbf{X}^{(2)} = dx^{(1)} dx^{(2)} \cos\theta, \tag{3.35}$$

where θ is the angle between $dx^{(1)}$ and $dx^{(2)}$. Since we are dealing with infinitesimal deformations, we expect θ to be near $\pi/2$ and instead employ the angle $\gamma \equiv \pi/2 - \theta$ and $\cos\theta = \sin\gamma \approx \gamma$. Assuming, as before, that $dx^{(1)} \approx d\mathbf{X}^{(1)}$ and $dx^{(2)} \approx d\mathbf{X}^{(2)}$, it follows that

$$\gamma \approx \frac{d\mathbf{X}^{(1)}}{dx^{(1)}} \cdot 2\boldsymbol{\varepsilon} \cdot \frac{d\mathbf{X}^{(2)}}{dx^{(2)}} = \hat{\mathbf{N}}_{(1)} \cdot 2\boldsymbol{\varepsilon} \cdot \hat{\mathbf{N}}_{(2)}. \tag{3.36}$$

If we select these unit vectors $\hat{\mathbf{N}}_{(1)}$ and $\hat{\mathbf{N}}_{(2)}$ to be aligned along any two different coordinate axes, say i and j, it follows then that

$$\gamma_{ij} = 2\varepsilon_{ij}, \tag{3.37}$$

defines the change in angle that has emerged between the two differential vectors. Given this interpretation, the off-diagonal components are sometimes referred to as the *engineering shear strain* components – the notion of shear being well-established by the non-zero angle γ_{ij}. In some engineering textbooks, the infinitesimal strain tensor is frequently written

$$[\varepsilon_{ij}] = \begin{pmatrix} \varepsilon_{11} & \frac{1}{2}\gamma_{12} & \frac{1}{2}\gamma_{13} \\ \frac{1}{2}\gamma_{21} & \varepsilon_{22} & \frac{1}{2}\gamma_{23} \\ \frac{1}{2}\gamma_{31} & \frac{1}{2}\gamma_{32} & \varepsilon_{33} \end{pmatrix}. \tag{3.38}$$

If $\hat{\mathbf{N}}_{(1)}$ and $\hat{\mathbf{N}}_{(2)}$ are oriented along principal strain directions, that γ_{12} is zero – an outcome of the fact that the deformed differential vectors retain the same orientation as their original form. Only the length has changed.

A differential vector aligned with the ith principal strain axis results in

$$dx^{(i)} = [1 + \varepsilon_i] \, d\mathbf{X}^{(i)}. \tag{3.39}$$

Thus, a differential volume $dX^{(1)} dX^{(2)} dX^{(3)}$, which has the form of a regular parallelepiped, retains that geometric form after transformation, but its volume has undergone a relative change

$$\frac{\Delta V}{V} = \frac{[1+\varepsilon_1] \, dX^{(1)} \, [1+\varepsilon_2] \, dX^{(2)} \, [1+\varepsilon_3] \, dX^{(3)}}{dX^{(1)} dX^{(2)} dX^{(3)}}$$
$$- \frac{dX^{(1)} dX^{(2)} dX^{(3)}}{dX^{(1)} dX^{(2)} dX^{(3)}}$$
$$\approx \varepsilon_1 + \varepsilon_2 + \varepsilon_3, \tag{3.40}$$

where we have neglected higher-order terms. We recognize, then, that the first invariant of the strain tensor describes the *cubical dilatation*, which is sometimes denoted by e. We observe, therefore, that $e = \Delta V/V = \varepsilon_{ii} = I_\varepsilon$. It is worthwhile noting here that differential volume change also implies differential density change. From mass conservation, it follows that $\rho V = (\rho + \Delta\rho)(V + \Delta V)$ thereby yielding $\Delta V/V = -\Delta\rho/\rho$. Volumetric change in Earth materials can emerge not only from changes in stress, but also from thermodynamic and structural changes. A prominent example in the Earth's interior relates to the magnesium iron silicates olivine and spinel (Carmichael, 1989). Olivine undergoes a phase transition into spinel deep in the mantle (Schubert *et al.*, 2001) and the accompanying $\Delta V/V \approx 10\%$ volumetric change may have an important role in deep focus earthquakes, where the epicenter resides 100s of kilometers below the surface.

Since the strain tensor has the same symmetric properties as the stress tensor, we can also develop Mohr's circles for small strain, as well as decompose the strain tensor into its spherical and deviator components. Recalling that the role of shear is to impart a small angular change into the deformed vectors, we employ in the construction of Mohr's circles the axial designations ε and $\gamma/2$ instead of σ_N and σ_S. Meanwhile, the *infinitesimal spherical strain tensor* is a diagonal matrix with equal elements denoted by $\varepsilon_M = \frac{1}{3}\varepsilon_{ii} = \frac{1}{3}e$, known as the *mean normal strain*. The *infinitesimal deviator strain tensor* $\boldsymbol{\eta}$ is defined by

$$\eta_{ij} = \varepsilon_{ij} - \frac{1}{3}\delta_{ij}\varepsilon_{kk} = \varepsilon_{ij} - \delta_{ij}\varepsilon_M. \tag{3.41}$$

The trace of $\boldsymbol{\eta}$ vanishes, and its principal strains are now $\eta_{(i)} = \varepsilon_{(i)} - \varepsilon_M$, and its principal strain directions remain unchanged.

Analogous to our previous discussion, a state of *plane strain* parallel to the $X_1 X_2$ plane exists if $\varepsilon_{33} = \gamma_{13} = \gamma_{31} = \gamma_{23} = \gamma_{32} = 0$. Similar explicit formulae exist for obtaining the remaining two nontrivial principal strain values.

Finally, we return to a theme we encountered at the beginning of this chapter in our discussion of Helmholtz's theorem on deformation. Once again, we consider two neighboring particles which had an initial displacement $d\mathbf{X}$ and now have a displacement $d\mathbf{x}$ as determined from the displacement differential

$$du_i = \frac{\partial u_i}{\partial X_j} dX_j. \tag{3.42}$$

It is convenient to express this displacement relative to the original differential $d\mathbf{X}$, that is

$$\frac{du_i}{dX} = \frac{\partial u_i}{\partial X_j} \frac{\partial X_j}{\partial X} = \frac{\partial u_i}{\partial X_j} N_j, \tag{3.43}$$

where N_j is now the unit vector in the direction of the original displacement. As before, we decompose this into its symmetric and anti-symmetric parts:

$$du_i = \left[\frac{1}{2}\left(\frac{\partial u_i}{\partial X_j} + \frac{\partial u_j}{\partial X_i}\right) + \frac{1}{2}\left(\frac{\partial u_i}{\partial X_j} - \frac{\partial u_j}{\partial X_i}\right)\right] dX_j = \left(\varepsilon_{ij} + \omega_{ij}\right) dX_j. \tag{3.44}$$

As before, ε_{ij} is the infinitesimal strain tensor, and

$$\omega_{ij} \equiv \frac{1}{2}\left(\frac{\partial u_i}{\partial X_j} - \frac{\partial u_j}{\partial X_i}\right) \tag{3.45}$$

is called the *infinitesimal rotation tensor*. As before, see especially section 1.5, we can define the *rotation vector* $\boldsymbol{\omega}$ by its components

$$\omega_i = -\epsilon_{ijk}\,\omega_{jk}, \tag{3.46}$$

which gives

$$du_i = \epsilon_{ijk}\,\omega_j\,dX_k + \varepsilon_{ij}\,dX_j. \tag{3.47}$$

In the case that there is no strain, we see that we can write

$$d\mathbf{u} = \boldsymbol{\omega} \times d\mathbf{X}, \tag{3.48}$$

which describes rigid-body rotation, as we saw in the first chapter.

3.4 Stretch ratios

It is useful to define the ratio of the magnitudes of $d\mathbf{x}$ and $d\mathbf{X}$ as the *stretch ratio*, Λ. For a differential element $d\mathbf{X}$ in the direction of the unit vector $\hat{\mathbf{N}}$, we define

$$\Lambda_{(\hat{\mathbf{N}})} = \frac{dx}{dX}, \tag{3.49}$$

although it is algebraically more convenient to deal with the square of this quantity. For convenience, we define $\mathbf{C}$ according to

$$C_{ij} = \frac{\partial x_k}{\partial X_i}\frac{\partial x_k}{\partial X_j}, \tag{3.50}$$

whereupon it follows that

$$(dx)^2 = d\mathbf{X} \cdot \mathbf{C} \cdot d\mathbf{X}, \tag{3.51}$$

so that dividing by $(dX)^2$

$$\Lambda^2_{(\hat{\mathbf{N}})} = \frac{d\mathbf{X}}{dX} \cdot \mathbf{C} \cdot \frac{d\mathbf{X}}{dX} = \hat{\mathbf{N}} \cdot \mathbf{C} \cdot \hat{\mathbf{N}}. \tag{3.52}$$

Importantly, this expression establishes that the relevant eigenvalues of **C** are positive or zero. In an analogous way, we define the stretch ratio $\lambda_{(\hat{\mathbf{n}})}$ according to the direction $\hat{\mathbf{n}} = d\mathbf{x}/dx$ by the equation

$$\frac{1}{\lambda_{(\hat{\mathbf{n}})}} = \frac{dX}{dx}. \tag{3.53}$$

For convenience, we define **c** according to

$$c_{ij} = \frac{\partial X_k}{\partial x_i} \frac{\partial X_k}{\partial x_j}, \tag{3.54}$$

and in a similar way obtain

$$\frac{1}{\lambda_{(\hat{\mathbf{n}})}^2} = \frac{d\mathbf{x}}{dx} \cdot \mathbf{c} \cdot \frac{d\mathbf{x}}{dx} = \hat{\mathbf{n}} \cdot \mathbf{c} \cdot \hat{\mathbf{n}}. \tag{3.55}$$

In general, $\Lambda_{(\hat{\mathbf{N}})} \neq \lambda_{(\hat{\mathbf{n}})}$ because $\hat{\mathbf{N}}$ is generally not the same as $\hat{\mathbf{n}}$. Similar considerations can be employed to calculate the angle between the deformed differential vectors.

As a final comment, it should be noted that the tensor **F** defined by

$$F_{ij} \equiv \frac{\partial x_i}{\partial X_j}, \tag{3.56}$$

owing to its invertibility, can be decomposed into one of two forms:

$$\mathbf{F} = \mathbf{R} \cdot \mathbf{U} = \mathbf{V} \cdot \mathbf{R}, \tag{3.57}$$

where **R** is an orthogonal rotation tensor and where **U** and **V** are symmetric, positive-definite tensors called the *right stretch tensor* and the *left stretch tensor*, respectively. As we showed in the first chapter, these forms represent the *polar decomposition* and have the physical meaning of converting the transformation process into a rotation followed by a stretching, or vice versa.

3.5 Velocity gradient

Suppose we know the velocity field $v_i = v_i(\mathbf{x}, t)$. We define the *spatial velocity gradient* according to

$$L_{ij} = \frac{\partial v_i}{\partial x_j}. \tag{3.58}$$

This tensor can be decomposed into its symmetric and anti-symmetric parts $L_{ij} = D_{ij} + W_{ij}$ according to the *rate of deformation tensor*

$$D_{ij} = \frac{1}{2}\left(\frac{\partial v_i}{\partial x_j} + \frac{\partial v_j}{\partial x_i}\right), \tag{3.59}$$

60   *Deformation and motion*

and the *vorticity* or *spin tensor*

$$W_{ij} = \frac{1}{2}\left(\frac{\partial v_i}{\partial x_j} - \frac{\partial v_j}{\partial x_i}\right). \tag{3.60}$$

We can write

$$\frac{\partial v_i}{\partial x_j} = \frac{\partial v_i}{\partial X_k}\frac{\partial X_k}{\partial x_j} = \frac{d}{dt}\left(\frac{\partial x_i}{\partial X_k}\right)\frac{\partial X_k}{\partial x_j}. \tag{3.61}$$

(This result uses the fact that material time derivatives and material gradients commute.) Our further discussion will be facilitated if we employ the tensor forms **L** and **F** defined by

$$L_{ij} = \frac{\partial v_i}{\partial x_j}; \quad \text{and} \quad F_{ij} = \frac{\partial x_i}{\partial X_j}, \tag{3.62}$$

wherein we now obtain

$$\mathbf{L} = \dot{\mathbf{F}} \cdot \mathbf{F}^{-1} \quad \text{or} \quad \dot{\mathbf{F}} = \mathbf{L} \cdot \mathbf{F}. \tag{3.63}$$

It is useful now to consider the evolution of the stretch ratio $\Lambda = dx/dX$ with respect to dX taken along $\hat{\mathbf{N}}$ which we introduced earlier. Recalling that $dx_i = (\partial x_i/\partial X_j)\, dX_j$ and the unit vectors $\hat{n}_i = dx_i/dx$ and $\hat{N}_j = dX_j/dX$, we may write

$$dx\, \hat{n}_i = \frac{\partial x_i}{\partial X_j} dX\, \hat{N}_j, \tag{3.64}$$

and obtain (by dividing by dX)

$$\hat{n}_i \Lambda = \frac{\partial x_i}{\partial X_j} \hat{N}_j \quad \text{or} \quad \hat{\mathbf{n}} \Lambda = \mathbf{F} \cdot \hat{\mathbf{N}}, \tag{3.65}$$

which describes both the rotation and stretching ratios. We now take the material derivative of this latter equation with respect to time:

$$\dot{\hat{\mathbf{n}}} \Lambda + \hat{\mathbf{n}} \dot{\Lambda} = \dot{\mathbf{F}} \cdot \hat{\mathbf{N}} = \mathbf{L} \cdot \mathbf{F} \cdot \hat{\mathbf{N}} = \mathbf{L} \cdot \hat{\mathbf{n}} \Lambda. \tag{3.66}$$

Dividing by Λ and taking the inner product with $\hat{\mathbf{n}}$ (recalling that $\hat{\mathbf{n}} \cdot \hat{\mathbf{n}} = 1$), we obtain

$$\hat{\mathbf{n}} \cdot \dot{\hat{\mathbf{n}}} + \dot{\Lambda}/\Lambda = \hat{\mathbf{n}} \cdot \mathbf{L} \cdot \hat{\mathbf{n}}. \tag{3.67}$$

Further, given that

$$\frac{d}{dt}(\hat{\mathbf{n}} \cdot \hat{\mathbf{n}}) = 2\hat{\mathbf{n}} \cdot \dot{\hat{\mathbf{n}}} = 0, \tag{3.68}$$

we obtain

$$\dot{\Lambda}/\Lambda = \hat{\mathbf{n}} \cdot \mathbf{L} \cdot \hat{\mathbf{n}} \quad \text{or} \quad \dot{\Lambda}/\Lambda = \frac{\partial v_i}{\partial x_j} n_i n_j. \tag{3.69}$$

This represents the *rate of stretching per unit stretch* of the element that originated in the $\hat{\mathbf{N}}$ direction and resulted in the $\hat{\mathbf{n}}$. This equation may be further simplified by eliminating the anti-symmetric portion of $\mathbf{L}$, namely

$$\dot{\Lambda}/\Lambda = D_{ij}\hat{n}_i\hat{n}_j. \tag{3.70}$$

The meaning of the diagonal elements is self-evident.

In order to appreciate the meaning of the off-diagonal terms, consider the evolution of two arbitrary deformations $d\mathbf{x}^{(1)}$ and $d\mathbf{x}^{(2)}$:

$$\frac{d}{dt}\left[dx_i^{(1)} dx_i^{(2)}\right] = \frac{d}{dt}\left[dx_i^{(1)}\right] dx_i^{(2)} + dx_i^{(1)} \frac{d}{dt}\left[dx_i^{(2)}\right]$$
$$= dv_i^{(1)} dx_i^{(2)} + dx_i^{(1)} dv_i^{(2)}. \tag{3.71}$$

Expanding dv_i as before, we obtain

$$\frac{d}{dt}\left[dx_i^{(1)} dx_i^{(2)}\right] = \left(\frac{\partial v_i}{\partial x_j} + \frac{\partial v_j}{\partial x_i}\right) dx_i^{(1)} dx_i^{(2)} = 2 D_{ij} dx_i^{(1)} dx_j^{(2)}. \tag{3.72}$$

Since the left-hand side is equal to $dx^{(1)} dx^{(2)} \cos\theta$, where θ defines the (time-dependent) angle between the two differentials, we now obtain

$$\frac{d}{dt}\left[dx^{(1)} dx^{(2)} \cos\theta\right] = \frac{d}{dt}\left[dx^{(1)} dx^{(2)}\right] \cos\theta - \left[dx^{(1)} dx^{(2)}\right] \sin\theta\, \dot{\theta}$$
$$= 2D_{ij} dx_i^{(1)} dx_j^{(2)}. \tag{3.73}$$

In the special case that $d\mathbf{x}^{(1)}$ is parallel to $d\mathbf{x}^{(2)}$, we recover (3.70). Suppose, on the other hand, that $d\mathbf{x}^{(1)}$ is perpendicular to $d\mathbf{x}^{(2)}$. Then, by (3.73),

$$2D_{ij}\hat{n}_i^{(1)}\hat{n}_j^{(2)} = 2\hat{\mathbf{n}}^{(1)} \cdot \mathbf{D} \cdot \hat{\mathbf{n}}^{(2)} = -\dot{\theta}. \tag{3.74}$$

This expression relates the evolution of the angle between the two originally orthogonal displacement vectors, thereby revealing the presence of shear through the *shear rate*. The ubiquitous factor of two, which we saw earlier in the derivation of γ, is pervasive in the engineering literature where it is customary to define the *rate of shear* as one half of the rate of change between two material line elements which are instantaneously at right angles.

3.6 Vorticity and material derivative

Recall that D_{ij} is a symmetric second-order tensor. Using this tensor, we can derive a number of different quantities, including principal values, principal directions, Mohr's circles representation, spherical and deviator tensors, and so on. In addition, it is possible to establish a connection between the material (time) derivative of the strain tensor in terms of **D**. We turn our attention now to **W**, namely

$$W_{ij} = \frac{1}{2}\left(\frac{\partial v_i}{\partial x_j} - \frac{\partial v_j}{\partial x_i}\right). \tag{3.75}$$

Further, we define a vorticity vector $\boldsymbol{\omega}$ defined by its components

$$\omega_i = \epsilon_{ijk}\,\partial_j\,v_k = -\epsilon_{ijk}\,W_{jk}. \tag{3.76}$$

(Note that this is the usual fluid mechanical definition of vorticity, which differs by a factor of 2 from engineering usage.) In particular, it follows that if D_{ij} vanishes, then

$$dv_i = \frac{1}{2}\epsilon_{ijk}\,\omega_j\,dx_k \quad \text{or} \quad d\mathbf{v} = \frac{1}{2}\boldsymbol{\omega}\times d\mathbf{x}. \tag{3.77}$$

The ambiguity associated with the factor of two can be resolved by considering a flow which manifests solid-body rotation at a rate $\boldsymbol{\Omega}$, namely $\mathbf{v} = \boldsymbol{\Omega}\times\mathbf{x}$. It follows that $\boldsymbol{\omega} \equiv \nabla\times\mathbf{v} = 2\boldsymbol{\Omega}$, the vorticity is twice the solid-body rotation rate.

We can summarize the physical interpretation of the velocity gradient **L** by noting that it separates the local instantaneous motion into a rate of stretching term D_{ij} and a rigid body rotation with angular velocity $\boldsymbol{\omega}$.

Finally, we turn our attention to the evolution of differential line elements. Recall that $F_{ij} = \partial x_i/\partial X_j$ and that $dx_i = F_{ij}\,dX_j$. We will define J as being the determinant of **F**. We define an infinitesimal surface element $d\mathbf{S}$ according to the two reference displacement vectors $d\mathbf{X}^{(1)}$ and $d\mathbf{X}^{(2)}$ according to

$$d\mathbf{S} = d\mathbf{X}^{(1)}\times d\mathbf{X}^{(2)} \quad \text{or} \quad dS_i = \epsilon_{ijk}\,dX_j^{(1)}\,dX_k^{(2)}. \tag{3.78}$$

Following the deformation, this is mapped into the new area element $d\mathbf{s}$ defined by

$$d\mathbf{s} = d\mathbf{x}^{(1)}\times d\mathbf{x}^{(2)} \quad \text{or} \quad ds_i = \epsilon_{ijk}\,dx_j^{(1)}\,dx_k^{(2)}. \tag{3.79}$$

We now wish to establish the connection between these two area elements.

We can immediately write

$$ds_i = \epsilon_{ijk}\,dx_j^{(1)}\,dx_k^{(2)} = \epsilon_{ijk}\,F_{jm}\,F_{kn}\,dX_m^{(1)}\,dX_n^{(2)}. \tag{3.80}$$

Recall, however, that

$$\epsilon_{ijk}\,F_{ip}\,F_{jm}\,F_{kn} = \epsilon_{pmn}\,J, \tag{3.81}$$

where J is the Jacobian of the transformation, which allows us to write

$$ds_i\,F_{ip} = \epsilon_{pmn}\,J\,dX_m^{(1)}\,dX_n^{(2)} = ds_i\,\frac{\partial x_i}{\partial X_p}. \tag{3.82}$$

Hence, it follows by introducing (3.78) that

$$ds_i \frac{\partial x_i}{\partial X_p} \frac{\partial X_p}{\partial x_q} = ds_q = \frac{\partial X_p}{\partial x_q} \, J \, dS_p. \tag{3.83}$$

The appearance of the Jacobian in this expression is an outcome of the usual volumetric transformation rules; the appearance of the $\partial X_p/\partial x_q$ term is an outcome of the need to rotate axes according to the deformation. Further details relating to Jacobians in vector calculus can be found in standard textbooks such as Arfken and Weber (2005), Millman and Parker (1977), and Schutz (1980).

Before proceeding further, it is useful to review some other properties of a general matrix **A**. In particular, we have already seen that

$$\det A = \epsilon_{ijk} \, A_{1i} \, A_{2j} \, A_{3k} = \epsilon_{ijk} \, A_{i1} \, A_{j2} \, A_{k3}, \tag{3.84}$$

and its generalization

$$\epsilon_{pqr} \det A = \epsilon_{ijk} \, A_{pi} \, A_{qj} \, A_{rk} = \epsilon_{ijk} \, A_{ip} \, A_{jq} \, A_{kr}. \tag{3.85}$$

Therefore, if we rearrange terms and interchange $\{ijk\} \leftrightarrow \{pqr\}$, we obtain

$$\epsilon_{ijk} \det A = A_{ip} \, \epsilon_{pqr} \, A_{jq} \, A_{kr}. \tag{3.86}$$

Suppose we select in the latter j and k so that $i \, j \, k$ are cyclic permutations of 123. Then, we observe that

$$\delta_{ik} = A_{ip} \, B_{pk}, \tag{3.87}$$

where **B** is defined according to

$$B_{pi} \equiv \frac{\epsilon_{pqr} \, A_{jq} \, A_{kr}}{\det(A)}, \tag{3.88}$$

where we are summing over q and r, but not over j and k, which are chosen to be cyclic with respect to i. We recognize that **B** is the inverse of **A** and that the summation reconstructs the "co-factor" of A_{ip}. We recognize this latter equation as an expression of *Cramer's Rule*. Helpful reviews of these results can be found in many textbooks including Boas (2006) and Arfken and Weber (2005). Differentiating (3.85) with respect to time, we obtain

$$\begin{aligned}\frac{d}{dt} \det A &= \epsilon_{ijk} \left(\dot{A}_{1i} \, A_{2j} \, A_{3k} + A_{1i} \, \dot{A}_{2j} \, A_{3k} + A_{1i} \, A_{2j} \, \dot{A}_{3k} \right) \\ &= \det A \left(\dot{A}_{1i} \, B_{ik} \, \delta_{k1} + \dot{A}_{2j} \, B_{jk} \, \delta_{k2} + \dot{A}_{3k} \, B_{k\ell} \, \delta_{\ell 3} \right) \\ &= \det A \, \operatorname{tr}\left(\dot{\mathbf{A}} \cdot \mathbf{B} \right). \end{aligned} \tag{3.89}$$

Now, recalling that J is the determinant of $\mathbf{F}$, it follows that

$$\dot{J} = \frac{d}{dt}(\det \mathbf{F}) = (\det \mathbf{F})\, \text{tr}\, (\dot{\mathbf{F}} \cdot \mathbf{F}^{-1}) = J\, \text{tr}\, \mathbf{L}. \tag{3.90}$$

Alternatively, this can be written

$$\dot{J} = J\, v_{i,i} = J\, \nabla \cdot \mathbf{v}. \tag{3.91}$$

Since J is the volume in a differential element, its time rate of change describes how the "density" of material contained within it changes – i.e. the density ρ' after the transformation satisfies $J \rho' = \rho$. Flows that are volume preserving are called incompressible or *isochoric* and satisfy $v_{i,i} = 0$. We will relate in a later chapter this expression to the continuity equation for the conservation of mass.

Recalling (3.84), we can write

$$d\mathbf{s} \cdot \mathbf{F} = J\, d\mathbf{S}, \tag{3.92}$$

which upon differentiating with respect to time, leads to

$$d\dot{\mathbf{s}} \cdot \mathbf{F} + d\mathbf{s} \cdot \dot{\mathbf{F}} = J\, (\text{tr}\, \mathbf{L})\, d\mathbf{S}. \tag{3.93}$$

Multiplying on the right by $\mathbf{F}^{-1}$, we obtain

$$d\dot{\mathbf{s}} + d\mathbf{s} \cdot \dot{\mathbf{F}} \cdot \mathbf{F}^{-1} = J\, (\text{tr}\, \mathbf{L})\, d\mathbf{S} \cdot \mathbf{F}^{-1}. \tag{3.94}$$

We now take the limit $\mathbf{s} \to \mathbf{S}$, i.e., take the limit $t \to 0$, so that

$$J\,(\text{tr}\, \mathbf{L})\, d\mathbf{S} \cdot \mathbf{F}^{-1} = (\text{tr}\, \mathbf{L})\, d\mathbf{S} = (\text{tr}\, \mathbf{L})\, d\mathbf{s}. \tag{3.95}$$

It then follows that

$$d\dot{\mathbf{s}} = (\text{tr}\, \mathbf{L})\, d\mathbf{s} - d\mathbf{s} \cdot \mathbf{L} \quad \text{or} \quad d\dot{s}_i = v_{k,k}\, ds_i - ds_j\, v_{j,i}. \tag{3.96}$$

Thus, we see how the rate of change of the element of area depends upon the flux of material into the volume, i.e. $v_{k,k}$, as well as the shear of the flow $v_{k,i}$. From these expressions, it follows that the deformation gradient $\mathbf{F}$ governs the stretch of a line element as well as the change in area and volume elements, while the velocity gradient $\mathbf{L}$ describes the *rate* at which these changes occur. Now that we have established a framework for the kinematics of deformation and motion, we turn our attention to the fundamental laws and equations for continua.

Exercises

3.1 Show, by direct evaluation using the chain rule, that

$$dx^2 = dX^2 + 2\, \varepsilon_{ik}\, dX_i\, dX_k,$$

and
$$dx^2 = dX^2 + 2\,e_{ik}\,dx_i\,dx_k,$$
where
$$\varepsilon_{ik} = \frac{1}{2}\left(\frac{\partial u_i}{\partial X_k} + \frac{\partial u_k}{\partial X_i} + \frac{\partial u_l}{\partial X_i}\frac{\partial u_l}{\partial X_k}\right),$$
and
$$e_{ik} = \frac{1}{2}\left(\frac{\partial u_i}{\partial x_k} + \frac{\partial u_k}{\partial x_i} - \frac{\partial u_l}{\partial x_i}\frac{\partial u_l}{\partial x_k}\right).$$

3.2 Suppose that $\mathbf{x}$ is defined in terms of $\mathbf{X}$ and t according to
$$x_1 = X_1\,\exp(t) + \cos(t) - 1,$$
$$x_2 = X_2\,\exp(-t) + 3\sin(2t),$$
and
$$x_3 = X_3 - \sin^2(t).$$
Describe the nature of this motion in terms of translation, rotation, and extension. (Explain the nature of translation, if any, the axis about which rotation takes place, if any, and the nature of extension. Show that the extension is not length preserving, but that volume elements are preserved.)

3.3 Suppose you are given $\mathbf{v}(\mathbf{x}, t)$. How would you go about calculating $\mathbf{x}(\mathbf{X}, t)$? Give a quantitative response (in terms of integrals, etc.) using the meaning of material derivative, etc. Hint: what is $\partial \mathbf{X}/\partial t$ where $\mathbf{x}$ is kept fixed?

3.4 Using the definitions in exercise 3.2, explicitly determine $\mathbf{v}(\mathbf{x}, t)$ and, treating those as differential equations, show by explicitly integrating the differential equations that you can recover the original definitions of the map. Do you believe that this inversion is formally possible for *any* $\mathbf{v}(\mathbf{x}, t)$? How is this related to the issue of "integrability" and to "chaos"? (Compare this with the Lorenz system shown in equations (6.1) in chapter 6.)

3.5 Calculate, using exercise 3.2, the associated deformation and velocity gradients. What does this tell you about the evolution of area and volume elements? Verify quantitatively the relationship between $\dot{J}$ and $v_{i,i}$ by evaluating J as a function of time and differentiating. Do the same for the evolution of differential area elements.

3.6 Consider the homogeneous deformation
$$x_1 = X_1 + \alpha\,X_2 + \alpha\beta\,X_3,$$
$$x_2 = \alpha\beta\,X_1 + X_2 + \beta^2\,X_3,$$

and
$$x_3 = X_1 + X_2 + X_3,$$
where α and β are constants. Show that
$$\beta = \frac{\alpha^2 + \alpha}{\alpha^2 + \alpha - 1},$$
is a sufficient condition for the map being volume preserving.

3.7 Suppose that $\det \mathbf{F} = 1$ for some deformation gradient. Under what circumstances does this map describe a rotation? Give your answer in *quantitative* and in *physical* terms. Hint: do not use the polar decomposition.

3.8 Assuming that Q_i is some vector, show that its flux through a surface S is given by
$$\frac{d}{dt}\int_S Q_i \, \hat{n}_i \, d^2x = \int_S \left(\dot{Q}_i + Q_i \, v_{k,k} - Q_k \, v_{i,k}\right) \hat{n}_i \, d^2x.$$

3.9 Verify the identity, where a_i is the acceleration vector,
$$\epsilon_{ijk} \, a_{k,j} = \dot{\omega}_i + \omega_i \, v_{j,j} - \omega_j \, v_{i kj}.$$
Using this result and that of the previous problem, show that
$$\frac{d}{dt}\int_S \omega_i \, \hat{n}_i \, d^2x = \int_S \epsilon_{ijk} \, a_{k,j} \, \hat{n}_i \, d^2x.$$

3.10 Suppose we have a velocity field given by
$$v_1 = a\,x_1 - b\,x_2, \qquad v_2 = b\,x_1 - a\,x_2, \qquad v_3 = c\sqrt{x_1^2 + x_2^2}.$$
Derive the nature of the density distribution that guarantees that the continuity equation is satisfied. Derive the consequent condition that guarantees that the flow is isochoric.

3.11 Consider the *homologous* expansion, which by definition has the form
$$x_i = f(t)\, X_i.$$
Assuming that the initial density is everywhere the same, derive ρ as a function of position and time, thereafter.

3.12 Show from the equations of motion and the continuity equation that
$$\frac{\partial(\rho v_i)}{\partial t} = \left(\sigma_{ij} - \rho\, v_i\, v_j\right)_{,j} + \rho\, b_i.$$
Show that the $\rho v_i v_j$ term is a form of stress and describes the rate of transform for momentum across a surface. Explain why it is sometimes called the "ram pressure."

4
Fundamental laws and equations

4.1 Terminology and material derivatives

We have now completed our derivation of the geometrical aspects of continuum mechanics. It is important to emphasize that our treatment is not complete, as we have restricted ourselves to a Cartesian description. We could have, albeit with much greater algebraic complexity, employed any curvilinear but mutually orthogonal coordinate system such as the family of systems known as *confocal quadrics* which include spherical, cylindrical, ellipsoidal, and many other (including Cartesian) coordinate systems. A brief discussion of these coordinate systems can be found in Arfken and Weber (2005) and an extensive exploration of all confocal quadrics is presented in Morse and Feshbach (1953). Had we done so, we would have had to introduce a "flat" or Cartesian basis, say $\mathcal{X}$, in addition to our reference coordinates $\mathbf{X}$ and transformed coordinates $\mathbf{x}$. In that case, we would have begun by defining an absolute measure of differential length ds according to the Pythagorean formula

$$ds^2 = d\mathcal{X}_k \, d\mathcal{X}_k, \tag{4.1}$$

from which we would derive

$$ds^2 = \frac{\partial \mathcal{X}_k}{\partial X_i} \frac{\partial \mathcal{X}_k}{\partial X_j} dX_i \, dX_j = g_{ij} \, dX_i \, dX_j, \tag{4.2}$$

where the so-called *metric tensor* g_{ij} is defined according to the partial derivatives in the intermediate expression. Owing to the increased complexity of the differential length, i.e. it is no longer simply $dX_i \, dX_i$, our previous derivations become much more complicated but nevertheless tractable. In particular, the metric g_{ij} and its derivatives are introduced in many of the expressions; the presence of derivative terms introduces the idea that curved geometry (cf. "flat space") has an important role. A familiar example of this is the Laplacian operator $\nabla^2 \equiv \partial_i \partial_i$ in flat space which becomes substantially more complicated, e.g. in cylindrical and spherical geometry. The role of curvature is the province of differential geometry and is included in some treatments of continuum mechanics. Since it is not necessary to use curvilinear coordinates, we will continue to employ Cartesian ones. In some cases, it may be more convenient to solve the resulting differential equations, e.g. those involving Laplacians and more complicated terms which can have some rather surprising forms, in different coordinate systems.

Conservation laws form the foundation of continuum mechanics, and we derive *all* laws from these conservation laws and, ultimately, variational principles. *Balance laws* are the simplest descriptions of these conservation laws and often have the form of a static balance. Energy balance, describing the interaction between mechanical and thermal energy, usually obeys a *virial theorem*. For example, in the important case of a harmonic oscillator potential, which is a good approximation describing the local molecular forces in a solid lattice, thermal and potential energies are in balance. When microscopic behavior is taken into account, the *second law of thermodynamics* or its statistical mechanical realization is of fundamental importance.

Virtually all of modern physics is predicated on the notion of variational principles. Hamilton's equations, Maxwell's equations, the fluid equations, even general relativity have a variational basis, although quantum mechanics is more difficult to treat. In some sense, the existence of a variational principle reflects the philosophy that nature always seeks the shortest path. By performing the necessary variation, we derive either static or dynamic equations describing the evolution of the physical variables and functions. These we refer to as *field equations*. Statistical mechanics is more problematic. If we assume that *entropy* is a maximum, as our variational principle, then we obtain the laws of equilibrium thermodynamics. We do not yet know in a general sense how to deal in statistical mechanics with non-equilibrium situations.

It is worth noting that computer methods for the solution of the field equations are based on this dichotomy of descriptions. *Finite difference* methods evolve in time, using differential versions of the time-dependent equations, the relevant field quantities. *Finite element* methods directly employ the variational principle by varying the field quantities until a suitable minimum is achieved. *Spectral* methods are a different representation of finite difference equations. Instead of looking at the point-by-point evolution of a field, we represent the field by some linear combination of basis set functions, e.g. sines and cosines. Then, we evolve using finite difference methods the amplitudes for the basis set functions over time. Another way to think of this, for the particular but important example of Fourier decomposition, is that we have replaced the role of the spatial representation by its Fourier transform and that we are using finite difference methods to solve the associated differential equations in the wavenumber representation.

Finally, *constitutive equations* are a description of the internal constitution and behavior of a material. It is important to note that this description is usually phenomenological as we do not usually have a direct description of material properties. To obtain that, it would be necessary to perform a statistical mechanical treatment of the underlying microscopic behavior and deduce from that the macroscopic behavior. This has rarely been done and most constitutive equations are not rigorously based but are approximate quantitative descriptions of what we observe. The inherent danger here emerges when we extrapolate the constitutive equations outside of the range of physical parameters for which the approximations are valid. An illustration of this emerges when we observe that for many solids the rate of change of the strain, neglecting tensor character for the moment, namely $\dot{\varepsilon}$, varies as a power law in the stress σ, or $\dot{\varepsilon} \propto \sigma^n$ where n is typically a number between 3 and 4. Clearly, this kind of expression must break down when molecular bonds are stretched beyond a certain length or other thermodynamic constraints are violated.

In the previous chapters, we derived some fundamental relations based solely on geometrical considerations relating to stress and to strain. In particular, recall that if

$$F_{ij} = \frac{\partial x_i}{\partial X_j}, \tag{4.3}$$

and, if J is the determinant of **F**, then

$$\dot{J} = v_{k,k} J. \tag{4.4}$$

Thus, suppose we consider the evolution of any scalar, vector, or tensor property, say $\mathcal{P}(t)$, of a collection of particles occupying a volume V, say

$$\mathcal{P}(t) = \int_V P(\mathbf{x}, t)\, d^3x, \tag{4.5}$$

where P represents the *density* associated with that quantity. As a practical example, $\mathcal{P}$ and P could represent the mass and mass density of an ensemble, although we are not restricted here to scalar quantities. Then it follows that

$$\frac{d}{dt}\mathcal{P}(t) = \frac{d}{dt}\int_V P(\mathbf{x}, t)\, d^3x = \frac{d}{dt}\int_V P\{\mathbf{x}(\mathbf{X}, t), t\}\, J\, d^3X. \tag{4.6}$$

We have made transition here to our (time-independent) reference coordinate **X** within the integral. Then, it follows,

$$\frac{d}{dt}\mathcal{P}(t) = \int_V \left(\dot{P} J + P \dot{J}\right) d^3X = \int_V \left(\dot{P} + v_{k,k} P\right) J\, d^3X. \tag{4.7}$$

Finally, we convert back to the spatial coordinate **x** to obtain

$$\frac{d}{dt}\mathcal{P}(t) = \int_V \left\{\dot{P}(\mathbf{x}, t) + v_{k,k} P(\mathbf{x}, t)\right\} d^3x. \tag{4.8}$$

It is important to note that the time derivative d/dt employed here is the material or total derivative

$$\frac{d}{dt} = \frac{\partial}{\partial t} + \mathbf{v} \cdot \nabla. \tag{4.9}$$

Hence, where for simplicity we have not explicitly shown the variable dependence for P, it follows that

$$\frac{d}{dt}\mathcal{P}(t) = \int_V \left\{ \frac{\partial P}{\partial t} + v_k \frac{\partial P}{\partial x_k} + v_{k,k} P \right\} d^3x. \tag{4.10}$$

We note that the last two terms in the integrand can be combined

$$\frac{d}{dt}\mathcal{P}(t) = \int_V \left\{ \frac{\partial P}{\partial t} + \frac{\partial}{\partial x_k}[v_k P] \right\} d^3x. \tag{4.11}$$

Note that the integrand now has the familiar form of the terms in the continuity equation. Employing Gauss' theorem, we obtain

$$\frac{d}{dt}\mathcal{P}(t) = \int_V \frac{\partial P}{\partial t} d^3x + \int_S v_k P \hat{n}_k d^2x, \tag{4.12}$$

where S describes the surface which bounds V. This latter result is often referred to as the *transport theorem*.

It is also useful to consider time derivatives of quantities integrated over some evolving surface. Consider $\mathcal{Q}(t)$ defined according to

$$\mathcal{Q}(t) = \int_S Q(\mathbf{x}, t) \, ds_p = \int_S Q(\mathbf{x}, t) \hat{n}_p \, d^2x, \tag{4.13}$$

where $Q(\mathbf{x}, t)$ is the surface density of that quantity. Here we are evaluating the projection of $\mathcal{Q}$ according to one specified direction. Recall from the previous chapter that

$$d\dot{s}_i = v_{k,k} \, ds_i - ds_k \, v_{k,i}. \tag{4.14}$$

Hence, it follows that

$$\frac{d}{dt}\mathcal{Q}(t) = \int_S (\dot{Q} + v_{k,k} Q) \, dS_p - \int_S Q \, v_{q,p} \, dS_q, \tag{4.15}$$

or, adapting to more conventional notation,

$$\frac{d}{dt}\mathcal{Q}(t) = \int_S (\dot{Q} + v_{k,k} Q) \hat{n}_p \, d^2x - \int_S v_{q,p} Q \hat{n}_q \, d^2x$$

$$= \int_S [(\dot{Q} + v_{k,k} Q)\delta_{pq} - v_{q,p} Q] \hat{n}_q \, d^2x. \tag{4.16}$$

Finally, we can examine the properties of a particle lying on the space curve C and expressed by the line integral

$$\mathcal{R}(t) = \int_C R(\mathbf{x}, t) \, dx_p. \tag{4.17}$$

We recall that

$$\frac{d}{dt}(dx_i) = v_{i,j} \, dx_j, \tag{4.18}$$

whereupon we obtain

$$\dot{\mathcal{R}}(t) = \int_C \dot{R} \, dx_p + \int_C v_{p,q} \, R \, dx_q = \int_C \left(\dot{R} \, \delta_{pq} + v_{p,q} \, R \right) dx_q. \tag{4.19}$$

This now completes our triad of material derivative expressions which we can now employ in dealing with physical problems.

4.2 Conservation of mass and the continuity equation

Suppose we consider any physical quantity whose totality is preserved, such as the mass or charge of a body. This notion can also be applied to more abstract quantities, such as the entropy or the total energy of a closed system, etc., and is not confined to mass alone. For convenience, we will refer to that conserved variable as mass in what follows, although it could be any physically conserved quantity. Consider a small volume ΔV which contains a mass Δm with reasonable mathematical continuity properties. Then we define the limit

$$\rho = \lim_{\Delta V \to 0} \frac{\Delta m}{\Delta V} \tag{4.20}$$

as the scalar field $\rho = \rho(\mathbf{x}, t)$ called the *mass density* at the position $\mathbf{x}$ at time t. Then, the total mass m of the body with volume V is given by

$$m = \int_V \rho(\mathbf{x}, t) \, d^3x. \tag{4.21}$$

This mass is assumed to be indistinguishable from the initial or reference mass with density field $\rho_0 = \rho_0(\mathbf{X}, t)$, namely

$$m = \int_{V^0} \rho_0(\mathbf{X}, t) \, d^3X, \tag{4.22}$$

where V^0 is the initial volume occupied by the configuration.

The *conservation of mass* law specifies that

$$\dot{m} = \frac{d}{dt} \int_V \rho(\mathbf{x}, t) \, d^3x = 0, \tag{4.23}$$

which becomes, using (4.10),

$$\dot{m} = \int_V \left[\frac{\partial \rho}{\partial t} + \frac{\partial}{\partial x_i} (\rho v_i) \right] d^3x = 0. \tag{4.24}$$

Since this equation must be satisfied independent of our choice of volume V, we obtain

$$\frac{\partial \rho}{\partial t} + \frac{\partial}{\partial x_i} (\rho v_i) = 0, \tag{4.25}$$

which we refer to as the *continuity equation* in Eulerian form. Alternatively, we can express this in the Lagrangian form

$$\frac{d\rho}{dt} + \rho v_{i,i} = 0; \tag{4.26}$$

Lagrangian forms for physical equations differ from the Eulerian form in most usage in that they employ the material derivative (i.e. the time derivative in a coordinate system instantaneously attached to the particles).

Another route to the same result emerges from the general properties of a transformation, something we alluded to in the previous chapter. In particular, we observe that

$$m = \int_V \rho(\mathbf{x}, t) \, d^3x = \int_{V^0} \rho_0(\mathbf{X}, t) \, d^3X$$

$$= \int_V \rho[\mathbf{x}(\mathbf{X}, t), t] \, d^3x = \int_{V^0} \rho(\mathbf{X}, t) \, J \, d^3X. \tag{4.27}$$

Considering the second and fourth steps, it follows that for arbitrary V^0

$$\int_{V^0} [\rho(\mathbf{X}, t) J - \rho_0(\mathbf{X}, t)] \, d^3X = 0. \tag{4.28}$$

Thus,

$$\rho J = \rho_0. \tag{4.29}$$

Since ρ_0 is time independent, we can also write the Lagrangian expression

$$\frac{d}{dt}[\rho J] = 0, \tag{4.30}$$

establishing a relationship we alluded to earlier.

This latter result is important when considering mass-specific quantities, e.g. quantities given in terms of some unit per unit mass, such as thermal energy. Denoting this mass-specific quantity $A(\mathbf{x}, t)$, we wish to obtain the rate of change of its integrated quantity, that is

$$\frac{d}{dt} \int_V A(\mathbf{x}, t) \, \rho \, d^3x = \frac{d}{dt} \int_{V^0} A(\mathbf{X}, t) \, \rho J \, d^3X = \int_{V^0} \dot{A}(\mathbf{X}, t) \, \rho J \, d^3X$$

4.3 Linear momentum and the equations of motion 73

$$= \int_{V^0} \dot{A}(\mathbf{x}, t) \, \rho \, d^3x. \tag{4.31}$$

The remarkable yet simple conservation law (4.31) has provided this fundamental relation, the continuity equation.

4.3 Linear momentum and the equations of motion

Suppose a continuum whose volume is presently V with bounding surface S is subjected to surface tractions $t_i^{(\hat{n})}$ and to distributed body forces $\rho \, b_j$. Unlike the situation we considered early in our treatment, we will now assume that the body is in motion with a velocity field $v_i = v_i(\mathbf{x}, t)$. We define the *linear momentum* of the body by

$$p_i = \int_V \rho \, v_i \, d^3x; \tag{4.32}$$

the *principle of linear momentum*, which is equivalent to Newton's second law, states that the time rate of change of the linear momentum is equal to the resultant force acting on the body. Expressing this in integral form, we obtain

$$\frac{d}{dt} \int_V \rho \, v_i \, d^3x = \int_V \rho \, \dot{v}_i \, d^3x = \int_S t_i^{(\hat{n})} \, d^2x + \int_V \rho \, b_i \, d^3x. \tag{4.33}$$

Recalling that

$$t_i^{(\hat{n})} = \sigma_{ji} \, \hat{n}_j, \tag{4.34}$$

and invoking Gauss' theorem, we obtain

$$\int_V (\rho \, \dot{v}_i - \sigma_{ji,j} - \rho \, b_i) \, d^3x = 0, \tag{4.35}$$

where we identify $\dot{v}_i$ as the *acceleration field* a_i of the body. Since V was arbitrary, we obtain

$$\rho \, \dot{v}_i = \sigma_{ji,j} + \rho \, b_i, \tag{4.36}$$

which are known as the *local equations of motion*. These equations are given in Lagrangian form, as the material derivative of the velocity is being employed. Since Euler was first responsible for deriving these expressions, they are often (confusingly) called the Euler equations of motion. On the other hand, the proper Eulerian form for these equations is

$$\rho \left[v_{i,t} + v_j \, v_{i,j} \right] = \sigma_{ji,j} + \rho \, b_i. \tag{4.37}$$

Note that the subscript ",t" is used as a shorthand for $\partial/\partial t$. In the special but important case $\dot{v}_i = 0$, we obtain the *equilibrium equations* of solid mechanics

$$0 = \sigma_{ji,j} + \rho\, b_i. \tag{4.38}$$

In fluid mechanics, the terms $v_j\, v_{i,j}$ in (4.37) are referred to as *inertial terms* and, for flows where the strain and shear rates are small, can often be ignored.

4.4 Piola–Kirchhoff stress tensor

We have defined the component σ_{ij} of the Cauchy stress tensor $\boldsymbol{\sigma}$ as the component in the x_j direction of the surface traction on a *deformed* surface element which is normal to the x_i direction in the current configuration, namely

$$t_j^{(\hat{\mathbf{n}})} = \sigma_{ij}\, \hat{n}_i. \tag{4.39}$$

For some purposes (Gurtin, 1981; Spencer, 1980; Mase and Mase, 1990), it is more convenient to use a stress tensor which is defined in terms of the traction on a material surface which is specified in the reference configuration. The symbol used most commonly for this tensor – which is called the *first Piola–Kirchhoff stress tensor* – is $\boldsymbol{\Pi}$ or, sometimes, $\boldsymbol{\Pi}^T$ as employed by Mase and Mase (1990). For most purposes, the Piola–Kirchhoff stress tensor is *not* employed because it mixes two coordinate systems – the current one and the reference one. To distinguish between these, we will use i as before to denote the current coordinate, but use α (with the use of Greek letters to denote "reference") to indicate the reference frame wherein traction is computed. Thus, we obtain

$$t_\alpha^{(\hat{\mathbf{n}})} = \Pi_{i\alpha}\, \hat{n}_i. \tag{4.40}$$

Thus, the benefit of this particular coordinate choice is that it makes the coordinate system wherein the stress (or traction) vector is calculated time-independent and, hence, not subject to change in expressions for the force, torque, and so on. This has the benefit of simplifying the Lagrangian expression for the equations of motion.

Further to our earlier discussion where we introduced continuity in the force and moment equations to develop the symmetry properties of the Cauchy stress tensor, we can perform similar manipulations with respect to the first Piola–Kirchhoff stress tensor. In particular, by transforming the first Piola–Kirchhoff stress tensor via the deformation tensor onto the current transformation – noting that this in general is not an orthogonal transformation – it is possible to show that the resulting matrix does preserve tensor transformation properties and is referred to as the *second Piola–Kirchhoff stress tensor*. It is sometimes encountered in consideration of the calculation of moments or couples; further, it can be related directly to the usual symmetric Cauchy stress tensor via the deformation tensor acting on each coordinate. However, it lacks any direct physical interpretation, and will not be discussed further.

Finally, it is worth mentioning that in the limit of small strain where linear theory applies, the Cauchy, Piola–Kirchhoff, and symmetric Piola–Kirchhoff stress tensors become formally equivalent.

4.5 Angular momentum principle

The *moment of momentum* is the moment of the linear momentum with respect to some point. The associated vector quantity is also called the *angular momentum* of the body. Of the nine possible moments, which might be considered as a matrix, only those associated with the anti-symmetric terms are generally considered. The trace of this matrix, which is related to what is called the "virial of Clausius" (Goldstein *et al.*, 2002), is helpful in a statistical mechanical description in relating the partitioning of kinetic and potential energies. The so-called "principle of angular momentum" states that the time rate of change of the moment of momentum of a body with respect to a given point is equal to the moment of the surface and body forces with respect to that point. Mathematically, this has the form

$$\frac{d}{dt}\int_V \epsilon_{ijk}\, x_j\, \rho\, v_k\, d^3x = \int_S \epsilon_{ijk}\, x_j\, t_k^{(\hat{n})}\, d^2x + \int_V \epsilon_{ijk}\, x_j\, \rho\, b_k\, d^3x. \tag{4.41}$$

Here, as before, we exploit the time-invariance of $\rho\, d^3x$ to obtain a simple expression for the left-hand side, namely $\int_V \epsilon_{ijk}\, x_j\, \rho\, \dot{v}_k\, d^3x$. Recalling the definition for the traction, we then obtain

$$\int_V \epsilon_{ijk}\left[x_j\left(\rho\, \dot{v}_k - \sigma_{qk,q} - \rho\, b_k\right) - \sigma_{jk}\right] d^3x = 0, \tag{4.42}$$

which reduces to

$$\int_V \epsilon_{ijk}\, \sigma_{kj}\, d^3x = 0 \tag{4.43}$$

for any volume. Thus, $\epsilon_{ijk}\, \sigma_{jk} = 0$ and, by direct expansion, it follows immediately that $\sigma_{jk} = \sigma_{kj}$. The derivation above assumes that no body or surface couples act on the body. If that is not the case and concentrated moments do apply, the material is said to be a *polar material* and the symmetry property of the Cauchy stress tensor no longer applies. This is a highly specialized situation which rarely emerges and will not be discussed further here.

4.6 Conservation of energy and the energy equation

The energy conservation law states that the material time derivative of the kinetic plus internal energies is equal to the rate of work of the sum of the surface and body forces, plus all other energies that enter or leave the body per unit time. These other energies can include, for example, mechanical, thermal, magnetic, or chemical energies. We will assume that we are dealing with a nonpolar continuum material (free of body or traction couples) and that only mechanical and thermal energies are significant. In our treatment, we are not including a substantial body of problems, including those characterized by phase transitions, dislocations in real materials, magnetic relaxation effects, and so on.

If only mechanical energy is considered, the energy balance can be derived from the equations of motion

$$\sigma_{ji,j} + \rho b_i = \rho \dot{v}_i. \tag{4.44}$$

In particular, consider the *kinetic energy* of a material occupying an arbitrary volume V of a body, namely

$$K(t) = \frac{1}{2} \int_V \rho \mathbf{v} \cdot \mathbf{v} \, d^3x = \frac{1}{2} \int_V \rho v_i v_i \, d^3x. \tag{4.45}$$

Mechanical power is defined by the *rate of work*, i.e. the rate at which a force acting over a distance applies. In the simplest situations, the mechanical power is simply $\mathcal{F} \cdot \mathbf{v}$, where $\mathcal{F}$ is the applied force and $\mathbf{v} \equiv d\mathbf{x}/dt$ is the applicable velocity. This immediately generalizes to the scalar $\mathcal{P}(t)$ defined by

$$\mathcal{P}(t) = \int_S t_i^{(\hat{n})} v_i \, d^2x + \int_V \rho b_i v_i \, d^3x. \tag{4.46}$$

Consider the material derivative of the kinetic energy integral:

$$\dot{K} = \frac{d}{dt} \int_V \frac{1}{2} \rho v_i v_i \, d^3x = \frac{1}{2} \int_V \rho \left[\frac{d}{dt} (v_i v_i) \right] d^3x$$

$$= \int_V \rho (v_i \dot{v}_i) \, d^3x = \int_V v_i (\sigma_{ji,j} + \rho b_i) \, d^3x. \tag{4.47}$$

Recalling that $v_i \sigma_{ij,j} = (v_i \sigma_{ij})_{,j} - v_{i,j} \sigma_{ij}$, we obtain

$$\dot{K}(t) = \int_V \left[\rho b_i v_i + (v_i \sigma_{ij})_{,j} - v_{i,j} \sigma_{ij} \right] d^3x; \tag{4.48}$$

we now use Gauss' theorem and the symmetrizing decomposition $v_{i,j} = D_{ij} + W_{ij}$ to write

$$\dot{K}(t) = \int_V \rho b_i v_i \, d^3x + \int_S t_i^{(\hat{n})} v_i \, d^2x - \int_V \sigma_{ij} D_{ij} \, d^3x. \tag{4.49}$$

4.6 Conservation of energy and the energy equation

Using (4.46), we then obtain

$$\dot{\mathcal{K}} + \mathcal{S} = \mathcal{P}, \tag{4.50}$$

where the integral

$$\mathcal{S} = \int_V \sigma_{ij} D_{ij} \, d^3x = \int_V \text{tr}\,(\boldsymbol{\Sigma} \cdot \mathbf{D}) \, d^3x \tag{4.51}$$

is known as the *stress work* and its integrand $\sigma_{ij} D_{ij}$ as the *stress power*. The balance of mechanical energy represented in (4.50) shows that of the total work done by external forces a portion goes toward increasing the kinetic energy and the remainder appears as work done by the internal stresses.

In general, $\mathcal{S}$ cannot be expressed as the material derivative of a volume integral, that is have the form, $\frac{d}{dt} \int_V (\ldots) \, d^3x$, because there is no universal function that we could insert as the integrand of this equation. In one special situation where

$$\mathcal{S} = \dot{\mathcal{U}} = \frac{d}{dt} \int_V \rho \, u \, d^3x = \int_V \rho \, \dot{u} \, d^3x, \tag{4.52}$$

where $\mathcal{U}$ is called the *internal energy* and u is the *specific internal energy* or *energy density* (per unit mass), equation (4.50) becomes

$$\frac{d}{dt} \int_V \left(\frac{1}{2} \rho \, v_i \, v_i + \rho \, u \right) d^3x = \int_V \rho \, b_i \, v_i \, d^3x + \int_S t_i^{(\hat{n})} v_i \, d^2x, \tag{4.53}$$

or simply

$$\dot{\mathcal{K}} + \dot{\mathcal{U}} = \mathcal{P}. \tag{4.54}$$

We employ u here as the specific energy density in conformity with standard practice; there should be no confusion with respect to the displacement $\mathbf{u}$. We observe, in the latter pair of equations, that part of the external work $\mathcal{P}$ causes an increase in kinetic energy, and the remainder is stored as internal energy, a feature shared by ideal elastic materials. Furthermore, this internal energy density u can be regarded on a microscopic level as the result of a micro-kinetic energy $\rho \, w_i \, w_i$ and a microscopic potential energy which binds individual atoms Φ_{pot}. Although we will not directly calculate quantities on this microscopic level, the physical imagery is important to our understanding of this phenomenon.

Finally, we consider thermal losses due to conduction in a *thermomechanical continuum*. The rate $\mathcal{Q}$ at which thermal energy is added to a body is represented by

$$\mathcal{Q} = \int_V \rho \, r \, d^3x - \int_S q_i \, \hat{n}_i \, d^2x. \tag{4.55}$$

Here, the scalar field r represents that rate at which heat per unit mass is produced or lost by internal sources, and is known as the *heat supply*. For example, radiative losses in a transparent medium is an example of this. Another is energy liberated, for example, by slow combustion. The vector q_i is called the *heat flux vector* and is a measure of the rate at which heat is conducted into the body per unit area per unit time across the element of surface d^2x whose outward normal is $\hat{n}_i$. The minus sign in the previous expression follows from heat flowing from hotter to colder regimes. The heat flux q_i is often assumed to obey *Fourier's law of heat conduction* (sometimes called "Fick's law")

$$q_i = -\kappa\, T_{,i} \quad \text{or} \quad \mathbf{q} = -\kappa\, \nabla T, \tag{4.56}$$

where κ is the *thermal conductivity* and $T_{,i}$ is the temperature gradient. This "law" is not universally obeyed; this is particularly problematic in metallic materials, such as in the earth's core.

With the addition of this thermal energy consideration, the complete energy balance requires modification of (4.54) to read

$$\dot{K} + \dot{U} = \mathcal{P} + \mathcal{Q}, \tag{4.57}$$

or, when expressed in detail,

$$\frac{d}{dt}\int_v \rho\left(\frac{1}{2} v_i\, v_i + u\right) d^3x = \int_V \rho\,(b_i\, v_i + r)\, d^3x + \int_S \left[t_i^{(\hat{n})} v_i - q_i\, \hat{n}_i\right] d^2x. \tag{4.58}$$

The latter may be regarded as a particular realization of the *second law of thermodynamics*. Converting the surface integral to a volume integral and utilizing the equations of motion, we then find the latter reduces to

$$0 = \int_V \left(\rho\, \dot{u} - \sigma_{ij}\, D_{ij} - \rho\, r + q_{i,i}\right) d^3x$$

$$= \int_V \left(\rho\, \dot{u} - \boldsymbol{\Sigma} : \mathbf{D} - \rho\, r + \nabla \cdot \mathbf{q}\right) d^3x, \tag{4.59}$$

or, briefly,

$$\mathcal{S} = \dot{U} - \mathcal{Q}; \tag{4.60}$$

this latter expression is sometimes referred to as the *thermal energy balance* (in analogy with our previously introduced mechanical energy balance) – however, this result lacks the absolute generality that the former one has. Since the above integral expressions hold for arbitrary volume, it follows that

$$\rho\, \dot{u} - \sigma_{ij}\, D_{ij} - \rho\, r + q_{i,i} = 0 \quad \text{or} \quad \rho\, \dot{u} - \boldsymbol{\Sigma} : \mathbf{D} - \rho\, r + \nabla \cdot \mathbf{q} = 0, \tag{4.61}$$

which is called the *energy equation*.

Summarizing, we observe that we have a direct mechanical energy balance equation which emerges from the equations of motion of linear momentum principle. With the addition of thermal energy, the balance becomes a (particular) statement of the first law of thermodynamics. Given that a first-principles treatment of all aspects of this problem is impractical, our discussion will be driven by empirical observations and phenomenological physics-based considerations. We will return to these themes later, in the context of entropy, temperature fields, and the second law of thermodynamics.

4.7 Constitutive equations

The preceding is a relatively general, within the limitations specified, description of the dynamic energy balance in continuum materials. The dependence of this picture upon their internal mechanical, thermal, and other properties is described by the so-called *constitutive equations*; it is important to note that the latter are generally phenomenological in character, i.e. deriving from empirical descriptions of experimental evidence, not *ab initio* microscopic behavior. Consequently, constitutive equations are generally organized according to the class of properties or behavior enjoyed by materials rather than the materials themselves. For example, similar constitutive equations differing perhaps only in the numerical values assigned to various empirical coefficients would be used to describe unrelated materials which possess the elastic–plastic response. Thus, constitutive equations are descriptions of certain kinds of material behavior and less that of characteristic materials. Proper constitutive equations should have a variety of features. For example, such equations are generally independent of any particular coordinate system, and spatially homogeneous, albeit not necessarily isotropic, and be Galilean invariant, i.e. independent of uniform velocity frames of reference. Note that this latter property is not preserved in so-called "cellular automata" models of fluid media, a major shortcoming of the theory. If mechanical and thermal fields do not influence each other significantly, we speak of an *uncoupled thermomechanical system* in which the constitutive equations involve only kinematic (strains, strain-rates, velocities) and mechanical (stresses, stress-rates) variables. One particular example is an *isothermal* situation where temperatures remain fixed throughout the medium.

We enumerate some well-known constitutive equations below (Mase and Mase, 1990). They are almost as varied as the imaginations of the scientists and engineers who have constructed them, and many more possibilities exist.

(1) *Linear elastic solids* (assuming infinitesimal strains) satisfy

$$\sigma_{ij} = C_{ijkm}\,\varepsilon_{km}, \tag{4.62}$$

where the C_{ijkm} are the elastic constants representing the properties of the body. For *isotropic* behavior, the latter takes the special form, as we will shortly derive,

$$\sigma_{ij} = \lambda \delta_{ij}\, \varepsilon_{kk} + 2\mu\, \varepsilon_{ij}, \tag{4.63}$$

where the λ and μ are the "Lamé coefficients" and express the elastic properties of the material. The general form in (4.62) employing a fourth-rank tensor emerges in anisotropic media such as crystalline materials and in depositional environments such as stream beds where size "sorting" of particles has taken place.

(2) The *linear viscous fluid*

$$\tau_{ij} = K_{ijmn}\, D_{mn}, \tag{4.64}$$

where τ_{ij} is the shear stress in the fluid and the constants K_{ijmn} represent its viscous properties. For a *Newtonian* fluid (i.e. one whose properties are independent of the flow),

$$\tau_{ij} = \lambda^* \delta_{ij}\, D_{kk} + 2\mu^*\, D_{ij}, \tag{4.65}$$

where λ^* and μ^* are viscosity coefficients. Note that the linear viscous fluid and stress–strain equations closely mirror each other, underscoring the link (at least in the linear regime) between the two types of continua.

(3) The *plastic* stress–strain equation,

$$d\varepsilon_{ij}^P = S_{ij}\, d\lambda, \tag{4.66}$$

where $d\varepsilon_{ij}^P$ is the plastic strain *increment*, S_{ij} is the deviator stress (in contrast with the mean normal stress), and $d\lambda$ is a proportionality constant.

(4) *Linear viscoelastic differential-operator* equations,

$$\{P\}\, S_{ij} = 2\, \{Q\}\, \eta_{ij} \tag{4.67}$$

$$\sigma_{ii} = 3K\, \varepsilon_{ii}, \tag{4.68}$$

where $\{P\}$ and $\{Q\}$ are differential time operators of the form

$$\{P\} = \sum_{i=0}^{N} p_i\, \frac{\partial^i}{\partial t^i} \tag{4.69}$$

$$\{Q\} = \sum_{i=0}^{N} q_i\, \frac{\partial^i}{\partial t^i}, \tag{4.70}$$

and in which the coefficients p_i and q_i (which are not necessarily constants) represent the viscoelastic properties. Also, K is the *bulk modulus* here. This set of equations specifies separately the deviatoric and volumetric responses. In most instances, only two terms where $N = 1$ are employed. One can show, using local Taylor series expansions, that these phenomenologically based constitutive relations as well as the ones that follow maintain memory effects.

(5) *Linear viscoelastic integral* equations

$$\eta_{ij} = \int_0^t \psi_s \left(t - t'\right) \frac{\partial S_{ij}}{\partial t'} dt' \tag{4.71}$$

$$\varepsilon_{ij} = \int_0^t \psi_v \left(t - t'\right) \frac{\partial \sigma_{ij}}{\partial t'} dt', \tag{4.72}$$

where the properties are contained within ψ_s and ψ_v, the *shear* and *volumetric creep functions*, respectively.

Creep is a process viewed macroscopically as a uniform flow, but is usually the outcome of large numbers of disjoint microscopic events including microfracture and/or microslip.

It is important to observe that each of the preceding constitutive equations, which are phenomenological descriptions of observed behavior, are expressly *linear*. This assumed linearity may be the outcome of a low-order Taylor series describing observed behavior over limited ranges of physical conditions. Real situations, particularly those encountered in geophysics, generally represent a much broader range of prevailing conditions, and some nonlinearity in the constitutive equations can be expected. In the next section, we will provide a little thermodynamic basis for this.

To formulate a well-posed problem in continuum mechanics, we need the field equations together with the *equations of state*, and the constitutive equations. (The latter are related to, but generally not equivalent to the equations of state which require a detailed micro-physical description.) Moreover, we require *boundary conditions* which describe the physical constraints on the evolution of the system. Together, this description provides a well-posed description for the time-evolution of the system. This can be obtained either analytically or numerically. As an illustration, the important field equations in both indicial and symbolic notations are given below.

(1) The continuity equation

$$\frac{\partial \rho}{\partial t} + (\rho \, v_k)_{,k} = 0 \quad \text{or} \quad \frac{\partial \rho}{\partial t} + \nabla \cdot (\rho \, \mathbf{v}) = 0. \tag{4.73}$$

(2) The equations of motion

$$\rho \, \dot{v}_i = \sigma_{ji,j} + \rho \, b_i \quad \text{or} \quad \rho \, \dot{\mathbf{v}} = \nabla \cdot \mathbf{\Sigma} + \rho \, \mathbf{b}. \tag{4.74}$$

(3) The energy equation

$$\rho \dot{u} - \sigma_{ij} D_{ij} - \rho r + q_{i,i} = 0 \quad \text{or} \quad \rho \dot{u} - \boldsymbol{\Sigma} : \mathbf{D} - \rho r + \nabla \cdot \mathbf{q} = 0. \quad (4.75)$$

Here, we have assumed that the body forces b_i and the distributed heat sources r are prescribed. We employ these quantities together with the *five* independent equations given above to determine *fourteen* unknowns, namely ρ, v_i, σ_{ij}, q_i and u. In addition, in non-isothermal situations, the entropy S and the temperature T fields need to be considered. Thus, we require *eleven* additional equations for *closure* of the system. Six of these will be constitutive equations (relating σ_{ij} to ϵ_{ij}), three will be temperature–heat conduction equations (i.e. Fourier's law $q_i = -\kappa T_{,i}$), and the remaining two will be equations of state. If the mechanical and thermal fields are uncoupled and an isothermal situation prevails, the continuity equation together with the equations of motion and the six constitutive equations provide a well-posed system of ten kinematic–mechanical equations for the ten unknowns ρ, v_i, and σ_{ij}. We will consider some simple examples in the next chapter, but now turn our attention to thermodynamic issues.

4.8 Thermodynamic considerations

Thermodynamics plays an important role in continuum mechanics in the geosciences, particularly in the context of flows in earth materials where the temperature can undergo dramatic change. This principle is exemplified by the cover photograph for this text of Devils Tower National Monument in northeastern Wyoming in the United States. There the basalt columns that formed on the periphery of this igneous intrusion are likely the outcome of a competition between thermal conduction and heat loss, on one hand, and thermal contraction and the resulting stresses that are produced, on the other. Most textbooks, however, do not go into significant depth on thermodynamics issues and they are largely overlooked in mathematical treatments. Turcotte and Schubert (2002) and Schubert *et al.* (2001) pursue some of the thermodynamic issues, especially thermal conduction and diffusion, in Earth and planetary-science applications. Relatively few treatments of continuum mechanics from a physicist's perspective are in widespread use in America or Europe, but the one due to Landau *et al.* (1986) can be applied to the geosciences. Only a few engineering textbooks devote much time to this topic, possibly because temperature variations encountered in applications are modest compared with those seen in geologic environments. Fung (1965), however, provides a very rich treatment of a range of issues that confront engineers while Narasimhan (1993) provides a more theoretical treatment. We begin by briefly reviewing some fundamentals of thermodynamics.

4.8 Thermodynamic considerations

The familiar form of the first law of thermodynamics is simply

$$d\mathcal{E} = -P\,dV, \tag{4.76}$$

and, upon introducing (non-adiabatic) heat losses dQ, we obtain the simple form of the second law

$$d\mathcal{E} = dQ - P\,dV = T\,dS - P\,dV. \tag{4.77}$$

Here, $\mathcal{E}$ is the internal energy, P is the pressure and V is the volume, while T is the temperature and S is the entropy. We will now proceed to derive these relations from the more fundamental continuum mechanical ones. In light of these fundamental relations in the adiabatic case, let U_R be the work done by the internal stresses per unit volume – note that this definition differs from the mass-specific ones discussed earlier. Consider the work done in deforming a body according to a change δu_i in the displacement vector u_i. Assuming that no body forces exist, the force (per unit volume) $\partial \sigma_{ij}/\partial x_j$ multiplied by the displacement δu_i integrated over the volume V of the body must satisfy

$$\int_V \delta U_R\, d^3x = \int_V \frac{\partial \sigma_{ij}}{\partial x_j} \delta u_i\, d^3x. \tag{4.78}$$

(Strictly speaking, the unit volumes before and after the deformation should be distinguished, since they in general contain different amounts of matter. In order to preserve the simplicity of this otherwise complex discussion, we will not make this distinction.) Integrating by parts with respect to the jth coordinate, we now obtain

$$\int_V \delta U_R\, d^3x = \oint \sigma_{ij} \delta u_i\, \hat{n}_j\, d^2x - \int_V \sigma_{ij} \frac{\partial \delta u_i}{\partial x_j} d^3x. \tag{4.79}$$

By considering an infinite medium, which is not deformed at infinity, the surface of integration in the first integral extends to infinity, and given that $\sigma_{ij} = 0$ on the surface, the integral vanishes. The second integral can, by virtue of the symmetry of the stress tensor, be written

$$\begin{aligned}
\int_V \delta U_R\, d^3x &= -\frac{1}{2}\int_V \sigma_{ij}\left(\frac{\partial \delta u_i}{\partial x_j} + \frac{\partial \delta u_j}{\partial x_i}\right) d^3x \\
&= -\frac{1}{2}\int_V \sigma_{ij}\,\delta\left(\frac{\partial u_i}{\partial x_j} + \frac{\partial u_j}{\partial x_i}\right) d^3x \\
&= -\int_V \sigma_{ij}\,\delta\varepsilon_{ij}\, d^3x.
\end{aligned} \tag{4.80}$$

Here, we have made use of the strain tensor defined earlier in the small strain limit. Thus, we find

$$\delta U_R = -\sigma_{ij}\,\delta\varepsilon_{ij}, \tag{4.81}$$

which may be regarded as an expression of the first law. This differs from our earlier treatment which involved material time derivatives – here we are looking for the absolute change in internal energy.

If the deformation of the body is fairly small, it then returns to its initial undeformed state when the external forces causing the deformation cease to act. Such deformations are said to be *elastic*. For large deformations, the removal of the external forces does not result in the total disappearance of the deformation; a *residual deformation* remains so that the state of the body is not that which existed before the forces were applied. Such deformations are said to be *plastic*. Microscopically, plastic deformations have resulted in an irreversible transformation of the structure of the material. See figure 5.1 in the next chapter. The primary mechanism for this is failure or fracture on a microscopic level, a subtle point which is often overlooked. When we stretch a spring well beyond its linear response and find that it does not return to the same equilibrium point, what has happened is that some of the bonds between atoms in the spring have broken and either have not reattached or have reattached differently while the spring was strongly deformed. Had the bonds reattached as before, then there would be no perceptible change in the macroscopic behavior. In the geophysical context, plasticity relates to the mechanically irreversible damage that accumulates due to strong deformations; these so-called "damage zones" – often associated with earthquake faults – are literally ensembles of broken or fragmented rock extending over a large range of sizes, i.e. of individual fragments. In what follows, we shall consider only elastic deformations. In chapter 5, some consideration will be given to the question of irreversibility.

We have implicitly assumed above that the process of deformation occurs so slowly that thermodynamic equilibrium is established in the body at every instant in accordance with the external conditions. Physically, equilibrium is maintained if microscopic behavior can communicate energy at speeds that are faster than the associated deformation rate. Since microscopic or thermal behavior is associated with the thermal velocity of the material, of order $\sqrt{k_B T / m} \approx 1$ km/s, where k_B is the Boltzmann constant and m is a measure of the mean molecular mass of "lattice" points, this condition is almost always justified. The process will then be thermodynamically if not mechanically reversible.

In what follows, we shall take all thermodynamic quantities, such as the entropy S, the internal energy $\mathcal{E}$, and so on as being relative to the unit volume of the body, introducing the approximations described above, and not relative to unit mass as often is done in fluid mechanics. An infinitesimal change $d\mathcal{E}$ in the internal energy is equal to the difference between the heat acquired by the unit volume considered and the work dU_R done by the internal stresses. The amount of heat is, for a reversible process, TdS. Thus,

4.8 Thermodynamic considerations

$$d\mathcal{E} = T\,dS - dU_R, \tag{4.82}$$

where δU_R was derived above, thereby giving

$$d\mathcal{E} = T\,dS + \sigma_{ij}\,d\varepsilon_{ij}. \tag{4.83}$$

This is the fundamental thermodynamic relation for deformed bodies.

In *hydrostatic compression*, the stress tensor is given by

$$\sigma_{ij} = -P\,\delta_{ij}; \tag{4.84}$$

this expression is isotropic, i.e. uniform in all directions, and the minus sign is the outcome of the force being oriented inward, i.e. the force is directed from high pressure to low pressure. There, it follows that

$$\sigma_{ij}\,d\varepsilon_{ij} = -P\,\delta_{ij}\,d\varepsilon_{ij} = -P\,d\varepsilon_{ii}. \tag{4.85}$$

We have seen in chapter 2 that the trace of the strain matrix, one of the invariants, defines in the linear regime the relative volume change due to the deformation. If we consider unit volume, therefore, ε_{ii} is just the change in that volume and $d\varepsilon_{ii}$ is the volume element dV. Thus, the thermodynamic relation takes its usual form

$$d\mathcal{E} = T\,dS - P\,dV \tag{4.86}$$

as expected.

We introduce the (Helmholtz) free energy $\mathcal{F}$ according to

$$\mathcal{F} = \mathcal{E} - T\,S, \tag{4.87}$$

whereupon we obtain

$$d\mathcal{F} = -S\,dT + \sigma_{ij}\,d\varepsilon_{ij}. \tag{4.88}$$

Finally, we introduce a thermodynamic potential (or Gibbs free energy) Φ according to

$$\Phi = \mathcal{E} - T\,S - \sigma_{ij}\,\varepsilon_{ij} = \mathcal{F} - \sigma_{ij}\,\varepsilon_{ij}, \tag{4.89}$$

which is a generalization of the usual expression

$$\Phi = \mathcal{E} - T\,S + P\,V. \tag{4.90}$$

The Gibbs free energy is important in that it is continuous even across phase transitions. Therefore, we find

$$d\Phi = -S\,dT - \varepsilon_{ij}\,d\sigma_{ij}. \tag{4.91}$$

Note that the internal energy, the Helmholtz free energy, and the thermodynamic potential or Gibbs free energy have independent variables which are determined according to the quantity selected. This is related to the *Maxwell relations* of thermodynamics and is a form of Legendre transformation. A fourth energy density is required to complete this picture, namely the enthalpy

$$\mathcal{H} = \mathcal{E} + PV, \tag{4.92}$$

in the usual case or

$$\mathcal{H} = \mathcal{E} - \sigma_{ij}\,\varepsilon_{ij}, \tag{4.93}$$

in the more general one. These different energy densities are indispensable as knowledge of them allows us to determine *any* thermodynamic quantity. The enthalpy is rarely used, as the only new information that it provides relates to the entropy S. Also, one of the free energy quantities is in some sense redundant as it provides information that we already have. Hence, we usually think of the internal energy $\mathcal{E}$ and the (Helmholtz) free energy $\mathcal{F}$ as providing the two equations of state we alluded to earlier. For example,

$$\sigma_{ij} = \left(\frac{\partial \mathcal{E}}{\partial \varepsilon_{ij}}\right)_S = \left(\frac{\partial \mathcal{F}}{\partial \varepsilon_{ij}}\right)_T. \tag{4.94}$$

Similarly, we obtain the strain from

$$\varepsilon_{ij} = -\left(\frac{\partial \Phi}{\partial \sigma_{ij}}\right)_T. \tag{4.95}$$

We now turn our attention to linear elasticity and classical fluids.

Exercises

4.1 Determine the form which the equations of motion take if the stress components are isotropic, i.e.

$$\sigma_{ij} = -P\,\delta_{ij},$$

where

$$P = P(\mathbf{x}, t).$$

4.2 Let a material continuum have the constitutive equation (appropriate for a linear elastic or linear viscous medium)

$$\sigma_{ij} = \alpha\,\delta_{ij}\,D_{kk} + 2\beta\,D_{ij},$$

where α and β are constants and where the deformation tensor $\mathbf{D}$ is defined by

$$D_{ij} = \frac{1}{2}\left(\frac{\partial v_i}{\partial x_j} + \frac{\partial v_j}{\partial x_i}\right).$$

Determine the form which the equations of motion take in terms of the velocity derivatives for this material, i.e. calculate the form for $\rho\,\dot{v}_i$.

4.3 For a rigid body rotation about the origin, the velocity field may be expressed by

$$v_i = \epsilon_{ijk}\,\Omega_j\,x_k,$$

where Ω_j is the angular velocity vector. Show that for this situation the angular momentum principle is given by

$$M_i = \frac{d}{dt}\left(I_{ij}\,\Omega_j\right),$$

where M_i is the total moment about the origin of all surface and body forces and I_{ij} is the moment of inertia of the body defined by the tensor

$$I_{ij} = \int_V \rho\left(\delta_{ij}\,x_k\,x_k - x_i\,x_j\right) d^3x.$$

4.4 Show that, for a rigid body rotation about the origin, the kinetic energy integral reduces to the form given in rigid-body dynamics, namely

$$K = \frac{1}{2}\,I_{ij}\,\Omega_i\,\Omega_j,$$

where I_{ij} is the inertia tensor defined in the previous exercise.

4.5 Consider a continuum for which the stress is

$$\sigma_{ij} = -P\,\delta_{ij}$$

and which obeys the heat conduction law

$$q_i = -\kappa\,T_{,i}.$$

Show that for this medium the energy equation takes the form

$$\rho\,\dot{u} = -P\,v_{i,i} + \rho\,r + \kappa\,T_{,ii}.$$

4.6 Suppose we have isotropic elastic behavior, namely

$$\sigma_{ij} = \lambda\,\delta_{ij}\,\varepsilon_{kk} + 2\,\mu\,\varepsilon_{ij},$$

where λ and μ are the Lamé coefficients. Show that

$$\sigma_{ii} = (3\,\lambda + 2\,\mu)\,\varepsilon_{ii},$$

and, using this result, deduce that

$$\varepsilon_{ij} = \frac{1}{2\,\mu}\left(\sigma_{ij} - \frac{\lambda}{3\,\lambda + 2\,\mu}\,\delta_{ij}\,\sigma_{kk}\right).$$

4.7 For a Newtonian fluid, the constitutive equation is given by

$$\sigma_{ij} = -P\,\delta_{ij} + \tau_{ij}$$
$$= -P\,\delta_{ij} + \lambda^*\,\delta_{ij}\,D_{kk} + 2\,\mu^*\,D_{ij}.$$

By substituting this constitutive equation into the equations of motion, derive the equation

$$\rho\,\dot{v}_i = \rho\,b_i - P_{,i} + \left(\lambda^* + \mu^*\right) v_{j,ji} + \mu^*\,v_{i,jj}.$$

These equations are commonly used to simulate the flow of viscous fluids.

5
Linear elastic solids

5.1 Elasticity, Hooke's law, and free energy

Elastic behavior is characterized by the stress in a material being a unique function of the strain, and that the material will be restored to its original or "natural" shape upon removal of the applied forces. Otherwise, we refer to the material as *inelastic*. It is convenient to consider the stress as having a Taylor series representation in terms of the strain. If it is necessary in our expression for the stress to keep powers higher than the first, i.e. the *linear* case, we call the stress–strain relationship *nonlinear*. If the stress is not a unique function of the strain, for example the stress–strain relationship shows hysteresis as in the accompanying figure, we observe that some change must occur in the internal energy as the integral of $\sigma_{ij}\, d\varepsilon_{ij}$ over the evolution of the system will be non-zero. Indeed, the system does not return to its original starting point.

Apart from such situations, the elastic model offers an important methodology for investigating linear problems. Appealing to the Taylor-series argument described earlier, we express the constitutive equation for elastic behavior in the general form

$$\sigma = \mathcal{G}(\varepsilon), \tag{5.1}$$

where $\mathcal{G}$ is a symmetric tensor-valued function and ε is any of the various strain tensors introduced earlier. Now, we turn to considerations of the free energy to establish the simplest meaningful form for the constitutive equations.

Recall that the free energy $\mathcal{F} = \mathcal{E} - TS$ satisfies the differential relation

$$d\mathcal{F} = -S\, dT + \sigma_{ij}\, d\varepsilon_{ij}, \tag{5.2}$$

so that the stress can be determined according to the thermodynamic Maxwell relation

$$\sigma_{ij} = \left(\frac{\partial \mathcal{F}}{\partial \varepsilon_{ij}}\right)_T. \tag{5.3}$$

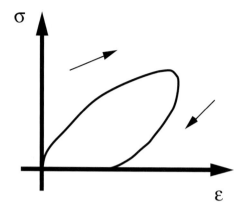

Figure 5.1 Nonlinear rheology and hysteresis.

It is natural to think of expanding the free energy in terms of powers of ε_{ij}. It is important to consider small deformation rates so that the temperature can be presumed constant for the present time throughout the body. Further, we will confine our attention to isotropic bodies, i.e. those not having any imposed internal structure. We imagine the undeformed state to be the state of the body in the absence of external forces and at the same temperature – the latter is necessitated by thermal expansion. Then, for $\varepsilon_{ij} = 0$, the internal stresses also vanish, i.e. $\sigma_{ij} = 0$. Thus, it follows that there is no linear term in the expansion of the free energy in powers of the strain.

Since the free energy is a scalar, each term in the expansion of $\mathcal{F}$ must be a scalar also. Only two independent scalars of the second degree can be constructed from the components of the symmetrical tensor ε_{ij}, namely the square of the sum, i.e. the trace, of the diagonal components $(\varepsilon_{ii})^2$ and the sum of the squares of all of the strain tensor's components $(\varepsilon_{ik})(\varepsilon_{ik})$. Expanding the free energy in powers of the strain tensor components, we therefore have for terms of the second order

$$\mathcal{F} = \mathcal{F}_0 + \frac{1}{2}\lambda\,(\varepsilon_{ii})^2 + \mu\,\varepsilon_{ij}\,\varepsilon_{ij} = \mathcal{F}_0 + \frac{1}{2}\lambda\,\varepsilon_{ii}^2 + \mu\,\varepsilon_{ij}^2. \tag{5.4}$$

This is the *general* expression[1] for the free energy of a deformed isotropic body; the quantities λ and μ are called *Lamé coefficients*.

In an earlier chapter, we observed that the relative change in volume in the deformation is given by the trace of the strain tensor ε_{ii}. Thus, if the trace vanishes, then

---

[1] We have also included the form of the expression often employed in other textbooks (Landau *et al.*, 1986). The key to understanding is to square the quantity under consideration.

5.1 Elasticity, Hooke's law, and free energy

the volume of the body is unchanged by the deformation, but its shape is altered. We refer to such a deformation as a *pure shear*. On the other hand, a deformation could cause a change in the volume of the body without changing its shape. In such a case, the deformation satisfies

$$\varepsilon_{ij} \propto \delta_{ij}. \tag{5.5}$$

The proportionality constant in the latter expression is a measure of the uniform relative expansion (or contraction) seen in each direction. We refer to such a deformation as being *hydrostatic*. Finally, we can represent any deformation as the sum of a pure shear and a hydrostatic compression by virtue of the identity

$$\varepsilon_{ij} = \left(\varepsilon_{ij} - \frac{1}{3}\delta_{ij}\,\varepsilon_{kk}\right) + \frac{1}{3}\delta_{ij}\,\varepsilon_{kk}. \tag{5.6}$$

This result may be regarded as an outcome of the Helmholtz theorem referred to earlier – with the translation term no longer present within the derivative of the displacement vector. The first term on the right is evidently a pure shear, since the sum of its diagonal terms vanishes, as a result of $\delta_{ii} = 3$. Meanwhile, the second term is isotropic and corresponds to a hydrostatic compression.

A convenient alternative to expression (5.4) for the free energy of a deformed isotropic body emerges if we exploit this decomposition into a pure shear and a hydrostatic compression. We will ignore here the constant term $\mathcal{F}_0$, which is the free energy of the undeformed body, as it has no influence on the stress of the body. We will reintroduce this term later when thermodynamic issues are introduced. We take as the two independent scalars of the second degree the sums of the squared components in (5.6). Then, $\mathcal{F}$ becomes

$$\begin{aligned}\mathcal{F} &= \mu\left(\varepsilon_{ij} - \frac{1}{3}\delta_{ij}\,\varepsilon_{kk}\right)\left(\varepsilon_{ij} - \frac{1}{3}\delta_{ij}\,\varepsilon_{kk}\right) + \frac{1}{2}K\,\varepsilon_{kk}\,\varepsilon_{kk} \\ &= \mu\left(\varepsilon_{ij} - \frac{1}{3}\delta_{ij}\,\varepsilon_{kk}\right)^2 + \frac{1}{2}K\,\varepsilon_{kk}^2.\end{aligned} \tag{5.7}$$

Let us compare the latter expression (5.7) with (5.4). We call the quantity K the *bulk modulus* (or *modulus of hydrostatic compression*) or *modulus of compression*, while μ is called the *shear modulus* or the *modulus of rigidity*. Comparing (5.4) with expression (5.7), we see that K is related to the Lamé coefficients by

$$K = \lambda + \frac{2}{3}\mu. \tag{5.8}$$

In thermodynamic equilibrium, the free energy is a minimum. If there are no external forces acting, then $\mathcal{F}$ must have a minimum for $\varepsilon_{ij} = 0$. Thus the quadratic form (5.8) physically must be positive. If the strain satisfies $\varepsilon_{ii} = 0$ (i.e. is incompressible), then the second term disappears in (5.8), while if the strain is hydrostatic (i.e. $\varepsilon_{ij} \propto \delta_{ij}$), then the first term vanishes. Hence, the necessary and sufficient condition for equation (5.8) to be positive is that each of the coefficients K and μ be positive (or zero). Thus, we conclude that the moduli of compression and rigidity are never negative:

$$K \geq 0, \quad \text{and} \quad \mu \geq 0. \tag{5.9}$$

In order to establish the stress tensor, it is necessary that we calculate the differential of the free energy. From (5.7), it follows that

$$d\mathcal{F} = K\,\varepsilon_{kk}\,d\varepsilon_{\ell\ell} + 2\mu\left(\varepsilon_{ij} - \frac{1}{3}\varepsilon_{kk}\,\delta_{ij}\right) d\left(\varepsilon_{ij} - \frac{1}{3}\varepsilon_{\ell\ell}\,\delta_{ij}\right). \tag{5.10}$$

In the second term, multiplication of the first parenthesis by δ_{ij} gives zero, leaving

$$d\mathcal{F} = K\varepsilon_{kk}\,d\varepsilon_{\ell\ell} + 2\mu\left(\varepsilon_{ij} - \frac{1}{3}\varepsilon_{\ell\ell}\,\delta_{ij}\right) d\varepsilon_{ij}, \tag{5.11}$$

which, upon writing $d\varepsilon_{ll} = \delta_{ij}\,d\varepsilon_{ij}$, gives

$$d\mathcal{F} = \left[K\,\varepsilon_{\ell\ell}\,\delta_{ij} + 2\mu\left(\varepsilon_{ij} - \frac{1}{3}\varepsilon_{\ell\ell}\,\delta_{ij}\right)\right] d\varepsilon_{ij}. \tag{5.12}$$

Hence, recalling (5.3), we obtain for the stress tensor

$$\sigma_{ij} = K\,\varepsilon_{kk}\,\delta_{ij} + 2\mu\left(\varepsilon_{ij} - \frac{1}{3}\delta_{ij}\,\varepsilon_{kk}\right). \tag{5.13}$$

This expression determines the stress tensor in terms of the strain tensor for an isotropic body. It shows that if the deformation is a pure shear ($\varepsilon_{ii} = 0$) or a pure hydrostatic compression ($\varepsilon_{ij} \propto \delta_{ij}$) then the stress–strain relation is determined only by the modulus of rigidity or of hydrostatic compression, respectively.

In order to obtain the strain in terms of the stress, we proceed in two steps. First, we take the trace of the preceding equation and obtain that

$$\varepsilon_{ii} = \frac{\sigma_{ii}}{3K}. \tag{5.14}$$

Substituting this expression in (5.13), we then obtain

$$\varepsilon_{ij} = \frac{\delta_{ij}\,\sigma_{\ell\ell}}{9K} + \frac{\sigma_{ij} - \frac{1}{3}\delta_{ij}\,\sigma_{\ell\ell}}{2\mu}, \tag{5.15}$$

which gives the strain tensor in terms of the stress tensor.

It is important to develop an intuitive feeling for the bulk modulus K. Consider the hydrostatic compressive situation where $\sigma_{ij} = -P\,\delta_{ij}$. Then, by (5.14), we obtain

5.1 Elasticity, Hooke's law, and free energy

$$\varepsilon_{ii} = -\frac{P}{K}. \tag{5.16}$$

Since the deformations are small, we may regard both ε_{ii} and P as small quantities. Hence, the ratio ε_{ii}/P can be regarded as the relative volume change to the pressure (change) where temperature is held fixed, namely $(1/V)(\partial V/\partial P)_T$. Thus,

$$\frac{1}{K} = -\frac{1}{V}\left(\frac{\partial V}{\partial P}\right)_T. \tag{5.17}$$

The quantity $1/K$ is called the *coefficient of hydrostatic compression* or, simply, the *coefficient of compression*.

Equation (5.15) shows that the strain tensor is a linear function of the stress tensor. Put another way, the deformation is proportional to the applied forces. This situation for small forces is called *Hooke's law*. This constitutive equation remains remarkably accurate for many materials (rubber is a well-known exception) even somewhat beyond the range of elastic deformations. It is useful to think that anelastic effects introduce higher-order terms into the free energy which initially have only a modest effect for deformations that extend beyond the elastic regime.

The form for the stress that we have obtained together with the algebraic properties of the free energy allow us to obtain a much simpler form for the free energy. In particular, we observe that

$$\varepsilon_{ij}\frac{\partial \mathcal{F}}{\partial \varepsilon_{ij}} = 2\mathcal{F}, \tag{5.18}$$

which we can verify analytically or regard as a consequence of the fundamentally quadratic form assumed and of "Euler's theorem" for homogeneous functions. Then, by virtue of $\partial \mathcal{F}/\partial \varepsilon_{ij} = \sigma_{ij}$, we obtain

$$\mathcal{F} = \frac{1}{2}\sigma_{ij}\,\varepsilon_{ij}. \tag{5.19}$$

Our discussion, thus far, has been motivated largely by thermodynamic and isotropy considerations. In engineering practice it is common to seek a general constitutive equation for (5.1) in the form of the linear (elastic) expression

$$\sigma_{ij} = C_{ijkm}\,\varepsilon_{km}, \tag{5.20}$$

where the tensor of elastic coefficients C_{ijkm} has $3^4 = 81$ components. The symmetry of the stress and strain relations, so that each has 6 independent terms, reduces the $81 = 9^2$ components to $36 = 6^2$ distinct coefficients. C_{ijkm} is a fourth-rank tensor and has tensor transformation properties in each of its indices (owing to the transformation properties of the stress and strain tensors). There is essentially no "physics" in this general linear model, but it does allow for behavior not contained within the thermodynamically derived relations. Earlier, we observed in passing that this could occur when dealing with crystalline materials or, in certain geologic contexts, situations where particle sizes in a material were "sorted" by gravity in depositional settings. In particular, if the body under consideration maintains some fundamental departure from homogeneity or anisotropy, it becomes necessary to invoke a more complete constitutive equation of the form (5.20). It is possible for C_{ijkm} to have temperature dependence, but as before we presume either *isothermal* (constant temperature) or *adiabatic* (no heat gain or loss). Further, we ignore strain–rate effects (i.e. we imagine the changes occurring sufficiently slowly so that thermodynamic equilibrium is maintained) and assume that the material under consideration is *homogeneous* so that the elastic coefficients remain unchanged. Owing to its absolute generality and linearity, we now refer to (5.20) as the *generalized Hooke's law*.

5.2 Homogeneous deformations

Situations wherein the strain tensor is constant throughout the volume of the body are called *homogeneous deformations*. Hydrostatic compression is an example of a homogeneous deformation. A further example is a *simple extension* (or compression) of a rod. We shall assume that the rod is aligned along the z-axis and allow forces to be applied to its ends which stretch it in both directions. We assume that these forces act uniformly over the end surface of the rod, and that the force per unit area is P.

Since the deformation is homogeneous, i.e. ε_{ij} is constant through the body, the stress tensor σ_{ij} is also constant, so it can be determined at once from the traction which now has the form of boundary conditions

$$\sigma_{ij}\hat{n}_j = P_i, \tag{5.21}$$

where the unit vector $\hat{n}_j$ is normal to the (local) surface. Here, the vector P_i is just a realization of the traction we referred to earlier and acts solely in the z-direction. There is no external force on the sides of the rod, and therefore

$$\sigma_{ij}\hat{n}_j = 0 \quad \text{for} \quad i = 1, 2. \tag{5.22}$$

Since the unit vector $\hat{\mathbf{n}}$ on the side of the rod is perpendicular to the z-axis, i.e. $\hat{n}_3 = 0$, it follows that all of the components of σ_{ij} except σ_{33} are zero. On the end surface, we have

$$\sigma_{3i}\hat{n}_i = P \quad \Rightarrow \quad \sigma_{33} = P. \tag{5.23}$$

We can now employ (5.15),

$$\varepsilon_{ij} = \frac{\delta_{ij}\sigma_{\ell\ell}}{9K} + \frac{\sigma_{ij} - \frac{1}{3}\delta_{ij}\sigma_{\ell\ell}}{2\mu}, \tag{5.24}$$

to evaluate the strain. In particular, it follows that all ε_{ij} for $i \neq j$ are necessarily zero. Meanwhile, the remaining components reduce to

$$\varepsilon_{11} = \varepsilon_{22} = -\frac{1}{3}\left(\frac{1}{2\mu} - \frac{1}{3K}\right)P; \quad \text{and} \quad \varepsilon_{33} = \frac{1}{3}\left(\frac{1}{3K} + \frac{1}{\mu}\right)P. \tag{5.25}$$

Note that the component ε_{33} gives the relative lengthening of the rod. The coefficient appearing before P is called the *coefficient of extension* and its reciprocal is called the *modulus of extension* or *Young's modulus*, E:

$$\varepsilon_{33} = \frac{P}{E}, \tag{5.26}$$

where

$$E = \frac{9K\mu}{3K + \mu}. \tag{5.27}$$

Meanwhile, the components ε_{11} and ε_{22} give the relative compression of the rod in the transverse direction. The ratio of the transverse compression to the longitudinal extension is called *Poisson's ratio* σ. Note that the use in the literature of σ to denote both Poisson's ratio and the components of the stress tensor should cause no ambiguity as the latter always has two indices. We find, then, that

$$\varepsilon_{11} \equiv \varepsilon_{22} \equiv -\sigma\,\varepsilon_{33}, \tag{5.28}$$

where

$$\sigma = \frac{1}{2} \cdot \frac{3K - 2\mu}{3K + \mu}. \tag{5.29}$$

Since the bulk modulus K and the Lamé coefficient μ are always positive, it follows that Poisson's ratio can vary between -1 (for $K = 0$) and $\frac{1}{2}$ (for $\mu = 0$). Thus,

$$-1 \leq \sigma \leq \frac{1}{2}. \tag{5.30}$$

In practice, Poisson's ratio varies only between 0 and $\frac{1}{2}$ and there is no substance known in nature where $\sigma < 0$; moreover, $\sigma > 0$ corresponds to the Lamé coefficient $\lambda > 0$ which would assure that the free energy is positive, which is thermodynamically desirable but not essential. Values of σ close to one-half, e.g. for rubber, correspond to a modulus of rigidity μ which is small compared with the modulus of compression K. The free energy reduces to $\mathcal{F} = \frac{1}{2}\sigma_{33}\varepsilon_{33}$ which in turn reduces to

$$\mathcal{F} = \frac{P^2}{2E}. \tag{5.31}$$

It has become customary to employ Young's modulus E and the Poisson ratio σ instead of K and μ in constitutive relations. To do this, observe that the equation defining σ can be expressed in terms of K/μ while the equation for E can be written in terms of K and K/μ, from which K can be directly determined and then μ. Inverting these formulae gives

$$\mu = \frac{E}{2(1+\sigma)} \quad \text{and} \quad K = \frac{E}{3(1-2\sigma)}. \tag{5.32}$$

The free energy now becomes

$$\mathcal{F} = \frac{E}{2(1+\sigma)}\left(\varepsilon_{ij}^2 + \frac{\sigma}{1-2\sigma}\varepsilon_{kk}^2\right). \tag{5.33}$$

The stress tensor is given in terms of the strain tensor by

$$\sigma_{ij} = \frac{E}{1+\sigma}\left(\varepsilon_{ij} + \frac{\sigma}{1-2\sigma}\varepsilon_{kk}\delta_{ij}\right). \tag{5.34}$$

Conversely, we find for the strain tensor

$$\varepsilon_{ij} = \frac{(1+\sigma)\sigma_{ij} - \sigma\,\sigma_{kk}\delta_{ij}}{E}. \tag{5.35}$$

Owing to the frequent use of the latter two formulae, we give them explicitly in component form:

$$\sigma_{11} = \frac{E}{(1+\sigma)(1-2\sigma)}[(1-\sigma)\varepsilon_{11} + \sigma(\varepsilon_{22}+\varepsilon_{33})]$$

$$\sigma_{22} = \frac{E}{(1+\sigma)(1-2\sigma)}[(1-\sigma)\varepsilon_{22} + \sigma(\varepsilon_{11}+\varepsilon_{33})]$$

$$\sigma_{33} = \frac{E}{(1+\sigma)(1-2\sigma)}[(1-\sigma)\varepsilon_{33} + \sigma(\varepsilon_{11}+\varepsilon_{22})] \tag{5.36}$$

$$\sigma_{12} = \frac{E}{1+\sigma}\varepsilon_{12}, \quad \sigma_{13} = \frac{E}{1+\sigma}\varepsilon_{13}, \quad \sigma_{23} = \frac{E}{1+\sigma}\varepsilon_{23}.$$

Conversely, we obtain

$$\varepsilon_{11} = \frac{1}{E}[\sigma_{11} - \sigma(\sigma_{22}+\sigma_{33})]$$

$$\varepsilon_{22} = \frac{1}{E}[\sigma_{22} - \sigma(\sigma_{11}+\sigma_{33})]$$

$$\varepsilon_{33} = \frac{1}{E}[\sigma_{33} - \sigma(\sigma_{11}+\sigma_{22})] \tag{5.37}$$

$$\varepsilon_{12} = \frac{1+\sigma}{E}\sigma_{12}, \quad \varepsilon_{13} = \frac{1+\sigma}{E}\sigma_{13}, \quad \varepsilon_{23} = \frac{1+\sigma}{E}\sigma_{23}.$$

A particularly important application of the latter sets of formulae emerges when we consider the compression of a rod whose sides are fixed in such a way that they cannot move. The external forces which cause the compression of the rod are applied to its ends and act along its length, which we again take to be along the z-axis. Such a deformation is called a *unilateral compression*. Since the rod is deformed only in the z-direction, only the component ε_{33} of the strain tensor is non-zero. Hence, it follows that

$$\sigma_{11} = \sigma_{22} = \frac{E\,\sigma}{(1+\sigma)(1-2\sigma)}\varepsilon_{33}, \quad \sigma_{33} = \frac{E\,(1-\sigma)}{(1+\sigma)(1-2\sigma)}\varepsilon_{33}. \tag{5.38}$$

Once again, we denote the compressive force by P (that is, $\sigma_{33} = P$, which is negative for a compression) and observe

$$\varepsilon_{33} = \frac{P\,(1+\sigma)(1-2\sigma)}{E\,(1-\sigma)}. \tag{5.39}$$

Finally, the free energy of the rod is

$$\mathcal{F} = \frac{P^2\,(1+\sigma)(1-2\sigma)}{2\,E\,(1-\sigma)}. \tag{5.40}$$

Having explored the role of homogeneous transformations, we now want to explore the role of thermodynamics.

5.3 Role of temperature

We now wish to introduce into our discussion the role of temperature change, either as a result of the deformation process itself or, commonly, from external sources. We shall regard as the undeformed state the state of the body in the absence of external forces at some given temperature T_0. If the body is at a temperature T different from T_0, then, even if there are no external forces, it will in general be deformed on account of thermal expansion. In the expression for the free energy $\mathcal{F}(T)$, both linear as well as quadratic terms will appear in the strain tensor. This linear term will represent the lowest order influence from temperature as well as strain. From the components of the strain tensor, only one linear scalar quantity can be formed, the trace of the matrix. We shall also assume that the temperature change $T - T_0$ which accompanies the deformation is small so that low-order Taylor series expansions remain valid. It is also important to remember that variation in temperature can occur both with respect to time as well as spatially. We then suppose that the coefficient of ε_{ii} in the expansion of the free energy, which must vanish for $T = T_0$, is simply proportional to the difference $T - T_0$. Thus, we find the free energy to be, building on equation (5.7) where we now explicitly include the $\mathcal{F}_0$ term:

$$\mathcal{F}(T) = \mathcal{F}_0(T) - K\alpha(T-T_0)\varepsilon_{kk} + \mu\left(\varepsilon_{ij} - \frac{1}{3}\delta_{ij}\varepsilon_{kk}\right)^2 + \frac{1}{2}K\varepsilon_{kk}^2, \quad (5.41)$$

where the coefficient of $T - T_0$ has been written as $-K\alpha$. The quantities μ, K, and α are assumed to be constant here – allowing for their temperature dependence would give rise to higher-order terms. Differentiating $\mathcal{F}$ with respect to ε_{ij} gives the stress tensor

$$\sigma_{ij} = -K\alpha(T-T_0)\delta_{ij} + K\varepsilon_{kk}\delta_{ij} + 2\mu\left(\varepsilon_{ij} - \frac{1}{3}\delta_{ij}\varepsilon_{kk}\right). \quad (5.42)$$

The first term gives the additional stresses caused by the change in temperature. In free thermal expansion of the body, that is where external forces are absent, no internal stresses can exist. Equating the stress tensor to zero, we find that the strain tensor ε_{ij} is of the form *constant* $\times \delta_{ij}$ and that

$$\varepsilon_{kk} = \alpha(T-T_0). \quad (5.43)$$

Since ε_{kk} is the relative change in volume caused by the deformation, α is clearly the *thermal expansion coefficient* of the body.

Among the various thermodynamically induced types of deformation, isothermal and adiabatic deformations are of importance. (Thus far, all of our expressions remain general.) In isothermal deformations, the temperature of the body does not change and we must put $T = T_0$ in (5.41) and we refer to the associated K and μ as *isothermal moduli*.

A deformation is adiabatic if there is no exchange of heat between the various parts of the body, or between the body and its surrounding medium. The entropy S, which is the derivative of the free energy with respect to temperature $-\partial \mathcal{F}/\partial T$, remains constant. Differentiating (5.41), we have by including only first-order terms in the strain tensor

$$S(T) = S_0(T) + K\alpha\varepsilon_{kk}. \quad (5.44)$$

Thus, it follows that (where S is now constant)

$$S = -\frac{\partial \mathcal{F}_0}{\partial T} + K\alpha\varepsilon_{kk}, \quad (5.45)$$

which can be used to approximate the $\mathcal{F}_0$ term (which incorporates a $T - T_0$ term) using the trace of the strain tensor in the expression for the free energy. This, in turn, leads to

$$\sigma_{ij} = K_{\text{adiabat}}\varepsilon_{kk}\delta_{ij} + 2\mu\left(\varepsilon_{ij} - \frac{1}{3}\delta_{ij}\varepsilon_{kk}\right), \quad (5.46)$$

5.3 Role of temperature

with the same modulus of rigidity μ as before, but with a different modulus of compression K_{adiabat}. The relation between the adiabatic and isothermal moduli can now be determined directly using a standard thermodynamic formula which we will not derive here, namely

$$\left(\frac{\partial V}{\partial P}\right)_S = \left(\frac{\partial V}{\partial P}\right)_T + \frac{T\,(\partial V/\partial T)_P^2}{C_P}, \tag{5.47}$$

where C_P is the specific heat per unit volume at constant pressure. If V is taken to be the volume occupied by matter which, before the deformation, occupied a unit volume, the derivative $\partial V/\partial T$ and $\partial V/\partial P$ gives the relative volume changes in heating and compression, respectively. From the preceding expressions we have that

$$\left(\frac{\partial V}{\partial T}\right)_P = \alpha, \quad \left(\frac{\partial V}{\partial P}\right)_S = -\frac{1}{K_{\text{adiabat}}}, \quad \left(\frac{\partial V}{\partial P}\right)_T = -\frac{1}{K}. \tag{5.48}$$

Thus, we find the following relation between the adiabatic and isothermal moduli, namely

$$\frac{1}{K_{\text{adiabat}}} = \frac{1}{K} - \frac{T\alpha^2}{C_P}, \quad \mu_{\text{adiabat}} = \mu. \tag{5.49}$$

Some more algebra then gives for Young's modulus and Poisson's ratio

$$E_{\text{adiabat}} = \frac{E}{1 - E\,T\,\alpha^2/9\,C_P},$$
$$\sigma_{\text{adiabat}} = \frac{\sigma + E\,T\,\alpha^2/9\,C_P}{1 - E\,T\,\alpha^2/9\,C_P}. \tag{5.50}$$

In practice, the term $E\,T\,\alpha^2/C_P$ is usually small, and it is usually sufficient to write

$$E_{\text{adiabat}} = E + E^2\,T\,\alpha^2/9\,C_P,$$
$$\sigma_{\text{adiabat}} = \sigma + (1+\sigma)\,E\,T\,\alpha^2/9\,C_P. \tag{5.51}$$

We recall that the stress tensor for an isothermal deformation is given in terms of the derivatives of the free energy

$$\sigma_{ij} = \left(\frac{\partial \mathcal{F}}{\partial \varepsilon_{ij}}\right)_T; \tag{5.52}$$

similarly, the stress tensor for an adiabatic deformation can be expressed in terms of the internal energy

$$\sigma_{ij} = \left(\frac{\partial \mathcal{E}}{\partial \varepsilon_{ij}}\right)_S. \tag{5.53}$$

Accordingly, it follows (using homogeneity arguments similar to ones employed earlier) that the internal energy can be written

$$\mathcal{E} = \frac{1}{2} K_{\text{adiabat}} \varepsilon_{kk}^2 + \mu \left(\varepsilon_{ij} - \frac{1}{3} \varepsilon_{kk} \delta_{ij} \right)^2. \tag{5.54}$$

This concludes our introduction to thermal effects, an important ingredient in geoscience environments.

5.4 Elastic waves for isotropic bodies

Here, we will derive the equations of equilibrium, but provide as well a brief derivation of the dynamical (i.e. wave) equations that emerge. We recall the general force equation (where we are implicitly employing the symmetry in the stress tensor)

$$\sigma_{ij,j} + \rho\, b_i = \rho\, \ddot{u}_i, \tag{5.55}$$

where we assume that the body forces b_i are known. Since we now have, from elasticity theory, constitutive equations relating the stress to the strain tensors, and recall that

$$\varepsilon_{ij} = \frac{1}{2} \left(\frac{\partial u_i}{\partial x_j} + \frac{\partial u_j}{\partial x_i} \right), \tag{5.56}$$

we now have a mechanism for exploring the evolution of the displacement vector and, hence, the strain. The evolution of the displacement vector is of paramount importance in seismology, as it describes the motion of the ground, for example, during an earthquake. Our focus in this section will be on static or equilibrium descriptions where the time dependence and external body forces are ignored. However, at the end, we will comment on how wave-like equations for longitudinal and transverse motions emerge.

Recalling the general expression (5.34) for the stress tensor in terms of Young's modulus and Poisson's ratio, we have

$$\frac{\partial \sigma_{ij}}{\partial x_j} = \frac{E\, \sigma}{(1+\sigma)(1-2\sigma)} \frac{\partial \varepsilon_{kk}}{\partial x_i} + \frac{E}{1+\sigma} \frac{\partial \varepsilon_{ij}}{\partial x_j}. \tag{5.57}$$

Introducing explicitly the relationship between the displacement vector and the strain tensor, we directly obtain the equations of equilibrium in the form

$$\frac{E}{2(1+\sigma)} \frac{\partial^2 u_i}{\partial x_j^2} + \frac{E}{2(1+\sigma)(1-2\sigma)} \frac{\partial^2 u_j}{\partial x_i \partial x_j} + \rho\, b_i = \rho\, \ddot{u}_i. \tag{5.58}$$

This equation describes *elastic oscillations* or *elastic waves* and it is instructive to consider a simple example of this phenomenon before proceeding to look at its more general properties.

In particular, consider a deformation **u** which is a function only of one spatial coordinate, say x, and of time, so that all derivatives with respect to y and to z vanish. We then obtain the wave equations, where we have set $b_i = 0$,

$$\frac{\partial^2 u_x}{\partial x^2} = \frac{1}{c_l^2} \frac{\partial^2 u_x}{\partial t^2}, \quad \frac{\partial^2 u_y}{\partial x^2} = \frac{1}{c_t^2} \frac{\partial^2 u_y}{\partial t^2}, \quad \frac{\partial^2 u_z}{\partial x^2} = \frac{1}{c_t^2} \frac{\partial^2 u_z}{\partial t^2}, \quad (5.59)$$

where the wave speeds are defined by

$$c_l = \sqrt{\frac{E(1-\sigma)}{\rho (1+\sigma)(1-2\sigma)}}; \quad c_t = \sqrt{\frac{E}{2\rho(1+\sigma)}}. \quad (5.60)$$

We have now defined three wave equations in one dimension where the quantities c_l and c_t are velocities of propagation of the wave. Note, in particular, that the speed of propagation of the component of deformation that is aligned along the axis of propagation, i.e. the longitudinal propagation along the x-axis, is greater than the speed of propagation of the transverse modes of deformation;

$$c_l > c_t. \quad (5.61)$$

It follows that the longitudinal mode is compressional in character, since the associated deformation is in the direction of propagation, but the meaning of the transverse mode requires a vectorial treatment.

Here, vector notation is particularly helpful. We observe that

$$\nabla \cdot \mathbf{u} \equiv \frac{\partial u_i}{\partial x_i} \quad \text{and} \quad \nabla^2 \mathbf{u} = \hat{e}_i \frac{\partial^2 u_i}{\partial x_j^2}. \quad (5.62)$$

Thus, the governing equations become

$$\nabla^2 \mathbf{u} + \frac{1}{1-2\sigma} \nabla (\nabla \cdot \mathbf{u}) = [\rho \ddot{\mathbf{u}} - \rho \mathbf{b}] \frac{2(1+\sigma)}{E}. \quad (5.63)$$

The equilibrium conditions emerge by ignoring the time-derivative terms. It is sometimes useful to employ the vector identity, for any vector function **v**,

$$\nabla (\nabla \cdot \mathbf{v}) = \nabla^2 \mathbf{v} + \nabla \times (\nabla \times \mathbf{v}). \quad (5.64)$$

Then, the governing equations become, where we now introduce c_l and c_t,

$$c_l^2 \nabla (\nabla \cdot \mathbf{u}) - c_t^2 \nabla \times (\nabla \times \mathbf{u}) = \ddot{\mathbf{u}} - \mathbf{b}. \quad (5.65)$$

We now want to explore how we can exploit the differences between the two operators in this equation.

5.5 Helmholtz's decomposition theorem

By taking the divergence of equation (5.65), the curl term disappears. We then obtain the equation for the evolution of $\nabla \cdot \mathbf{u}$, while taking the curl of this equation, the gradient term disappears and we obtain the equation for the evolution of $\nabla \times \mathbf{u}$. It is useful to decompose the displacement vector $\mathbf{u}$ into its transverse and longitudinal components

$$\mathbf{u} = \mathbf{u}_t + \mathbf{u}_l, \tag{5.66}$$

where

$$\nabla \times \mathbf{u}_l = 0 \quad \text{and} \quad \nabla \cdot \mathbf{u}_t = 0. \tag{5.67}$$

This decomposition is sometimes referred to as *Helmholtz's Decomposition Theorem* (Arfken and Weber, 2005). This decomposition was first employed in electromagnetic theory (Jackson, 1999) in application to solenoidal currents and we will demonstrate that equation (5.66) holds by construction.

We begin by defining the two components according to

$$\mathbf{u}_t = -\frac{1}{4\pi} \nabla \int \frac{\nabla' \cdot \mathbf{u}(\mathbf{x}')}{|\mathbf{x} - \mathbf{x}'|} d^3 x'$$

$$\mathbf{u}_l = +\frac{1}{4\pi} \nabla \times \nabla \times \int \frac{\mathbf{u}(\mathbf{x}')}{|\mathbf{x} - \mathbf{x}'|} d^3 x', \tag{5.68}$$

where the operator ∇ is with respect to $\mathbf{x}$ and the operator ∇' is with respect to $\mathbf{x}'$ and where the integration is performed over all space in $\mathbf{x}'$. We can integrate by parts our expression for $\mathbf{u}_t$ and obtain

$$\begin{aligned}\mathbf{u}_t &= +\frac{1}{4\pi} \nabla \int \mathbf{u}(\mathbf{x}') \cdot \nabla' \left(\frac{1}{|\mathbf{x} - \mathbf{x}'|} \right) d^3 x' \\ &= -\frac{1}{4\pi} \nabla \int \mathbf{u}(\mathbf{x}') \cdot \nabla \left(\frac{1}{|\mathbf{x} - \mathbf{x}'|} \right) d^3 x',\end{aligned} \tag{5.69}$$

where we replaced inside the integral ∇' by ∇. This then yields

$$\mathbf{u}_t = -\frac{1}{4\pi} \nabla \left[\nabla \cdot \int \frac{\mathbf{u}(\mathbf{x}')}{|\mathbf{x} - \mathbf{x}'|} d^3 x' \right] \tag{5.70}$$

to which we can now add our expression for $\mathbf{u}_l$ and obtain

$$\mathbf{u}_t + \mathbf{u}_l = \frac{1}{4\pi}\nabla \times \nabla \times \int \frac{\mathbf{u}(\mathbf{x}')}{|\mathbf{x}-\mathbf{x}'|}d^3x' - \frac{1}{4\pi}\nabla\nabla \cdot \int \frac{\mathbf{u}(\mathbf{x}')}{|\mathbf{x}-\mathbf{x}'|}d^3x'$$

$$= -\frac{1}{4\pi}\nabla^2 \int \frac{\mathbf{u}(\mathbf{x}')}{|\mathbf{x}-\mathbf{x}'|}d^3x'. \tag{5.71}$$

Finally, we can now write

$$\mathbf{u}_t + \mathbf{u}_l = -\frac{1}{4\pi}\int \mathbf{u}(\mathbf{x}')\nabla'^2\left(\frac{1}{|\mathbf{x}'-\mathbf{x}|}\right)d^3x'. \tag{5.72}$$

In chapter 1, exercise 1.14, we showed by using Gauss' theorem that

$$\nabla'^2\left(\frac{1}{|\mathbf{x}'-\mathbf{x}|}\right) = 0 \tag{5.73}$$

so long as $\mathbf{x}' \neq \mathbf{x}$ but that the integral over all $\mathbf{x}'$ (with $\mathbf{x}$ fixed) was -4π. This showed that there was a singularity present in the former expression which, it turns out, is related to the *Dirac delta function* (Arfken and Weber, 2005; Mathews and Walker, 1970). This result, moreover, is intimately related to what are called *Green's functions*, which occupy an important role in more advanced applications of continuum mechanics. In terms of equation (5.72), we only need to integrate $\mathbf{x}'$ over an infinitesimal sphere containing $\mathbf{x}$, since there is no contribution to the integral from outside. Then we can exploit partial integration and Gauss' theorem whereupon the right-hand side of equation (5.72) reduces to $\mathbf{u}(\mathbf{x})$, thereby verifying that equation (5.66) is satisfied.

From this decomposition, we now obtain separate wave equations for the longitudinal and transverse motion:

$$c_l^2 \nabla^2 \mathbf{u}_l = \ddot{\mathbf{u}}_l - \mathbf{b}_l, \tag{5.74}$$

and

$$c_t^2 \nabla^2 \mathbf{u}_t = \ddot{\mathbf{u}}_t - \mathbf{b}_t. \tag{5.75}$$

We now observe that the transverse mode of propagation corresponds to incompressible behavior or to pure shear. Moreover, it is easy to show that a homogeneous deformation (or hydrostatic compression), i.e. one where $\varepsilon_{ij} \propto \delta_{ij}$, corresponds directly to the longitudinal mode of propagation. Hence, longitudinal and transverse modes of propagation are often referred to as compressional and shear waves, respectively. Moreover, owing to the inequality in wave speeds, they are also referred to as *P*, as in primary, and *S*, as in secondary, waves. The solution of these wave equations in different media with different Young's moduli and Poisson ratios is the province of seismology and we will not consider the role of time dependence further. There are many excellent sources available that further develop the theory and its applications, particularly in the context of seismology (Bullen and Bolt, 1985; Kennett, 1983; Ben-Menahem and Singh, 2000; Aki and Richards, 2002; Shearer, 2009).

5.6 Statics for isotropic bodies

A very important case is that where the deformation of the body is caused, not by body forces, but by forces applied to its surface. It is important here to note that, having derived the general albeit linear equations for equilibrium, we can now solve them in any coordinate system that we wish. For example, the Laplacian can be expressed using spherical coordinates; we would have arrived at the same result had we begun by using curvilinear coordinates instead of rectilinear ones and developed our deformation equations consistent with the curvature of the spatial coordinate system that we employed. We have taken the simpler route. What is important here is that we have the ability now to select the coordinate system that makes our calculation least complicated. The equations of equilibrium then become

$$(1 - 2\sigma)\nabla^2 \mathbf{u} + \nabla(\nabla \cdot \mathbf{u}) = 0, \tag{5.76}$$

or

$$2(1 - \sigma)\nabla(\nabla \cdot \mathbf{u}) - (1 - 2\sigma)\nabla \times (\nabla \times \mathbf{u}) = 0. \tag{5.77}$$

Now, the external forces appear in the solution only through the boundary conditions.

Taking the divergence of (5.76), and using the identity

$$\nabla \cdot \nabla = \nabla^2, \tag{5.78}$$

we find

$$\nabla^2 (\nabla \cdot \mathbf{u}) = 0. \tag{5.79}$$

Thus, $\nabla \cdot \mathbf{u}$, which is a measure of volumetric change due to deformation, is a harmonic function and can be expressed via linear combinations of basis set functions, such as spherical harmonics. Taking the Laplacian of (5.76), we obtain

$$\nabla^2 \nabla^2 \mathbf{u} = 0. \tag{5.80}$$

In equilibrium, we say that the displacement vector satisfies the *biharmonic equation* (5.80). In many sources, the operator ∇^2 is sometimes written Δ, and the biharmonic operator $\nabla^2 \nabla^2$ is expressed as $\Delta\Delta$ or Δ^2. These equations remain valid in a uniform gravitational field (since $\nabla \cdot \mathbf{u} = 0$ where the body force is the local gravitational acceleration $\mathbf{g}$), albeit not in the general case of external forces which vary through the body. Note however that the general solution of the equations of equilibrium in the absence of body forces cannot be an arbitrary biharmonic vector, as the displacement vector must also satisfy the lower-order equation (5.76).

If a body is non-uniformly heated, an additional term appears in the equation of equilibrium. In particular, we recall the general stress tensor

$$\sigma_{ij} = -K\alpha(T-T_0)\delta_{ij} + K\varepsilon_{kk}\delta_{ij} + 2\mu\left(\varepsilon_{ij} - \frac{1}{3}\delta_{ij}\varepsilon_{kk}\right). \quad (5.81)$$

Accordingly, $\partial\sigma_{ij}/\partial x_j$ contains a term

$$-K\alpha\frac{\partial T}{\partial x_i} = -\left[\frac{E\alpha}{3(1-2\sigma)}\right]\frac{\partial T}{\partial x_i}. \quad (5.82)$$

The equation of equilibrium now takes the form

$$\frac{3(1-\sigma)}{1+\sigma}\nabla(\nabla\cdot\mathbf{u}) - \frac{3(1-2\sigma)}{2(1+\sigma)}\nabla\times(\nabla\times\mathbf{u}) = \alpha\nabla T. \quad (5.83)$$

This equation offers some important insight into the role of thermal expansion and contraction. For example, one can verify by inspection that a uniform temperature gradient in a medium gives rise to a displacement which is oriented along the temperature field and which varies linearly with distance along the temperature gradient.

A popular class of problems in elasticity theory emerges when one component of the displacement vector, say u_z, is fixed at zero throughout the body. Then, all functional dependence resides within the x and y variables, and the theory of functions of a complex variable, including especially conformal mapping can be invoked. This provides a particularly powerful way of solving elasticity problems with complicated boundary conditions. As a simple illustration, recall that in the absence of external body forces $\partial\sigma_{ij}/\partial x_j = 0$ now reduces to two equations

$$\frac{\partial\sigma_{xx}}{\partial x} + \frac{\partial\sigma_{xy}}{\partial y} = 0, \quad \frac{\partial\sigma_{xy}}{\partial x} + \frac{\partial\sigma_{yy}}{\partial y} = 0. \quad (5.84)$$

A general solution to the latter has the form

$$\sigma_{xx} = \frac{\partial^2\chi}{\partial y^2}, \quad \sigma_{xy} = -\frac{\partial^2\chi}{\partial x\,\partial y}, \quad \sigma_{yy} = \frac{\partial^2\chi}{\partial x^2}, \quad (5.85)$$

where χ is an arbitrary function of x and y. A solution to this problem necessarily exists since the three quantities σ_{xx}, σ_{xy}, σ_{yy} can be expressed in terms of the two quantities u_x and u_y and, therefore, are not independent. Using our fundamental relationships (5.36) for the stress in terms of the strain, we obtain for a plane deformation

$$\sigma_{xx} + \sigma_{yy} = \frac{E\left(u_{x,x} + u_{y,y}\right)}{(1+\sigma)(1-2\sigma)}. \tag{5.86}$$

However, we have from above that

$$\sigma_{xx} + \sigma_{yy} = \nabla^2 \chi, \quad u_{x,x} + u_{y,y} = \frac{\partial u_x}{\partial x} + \frac{\partial u_y}{\partial y} \equiv \nabla \cdot \mathbf{u}. \tag{5.87}$$

However, we also established that $\nabla \cdot \mathbf{u}$ is harmonic, i.e. $\nabla^2 (\nabla \cdot \mathbf{u}) = 0$, which then leaves us with

$$\nabla^2 \nabla^2 \chi = 0; \tag{5.88}$$

that is, χ is biharmonic and is sometimes called the *stress function*. When the plane problem has been solved and the function χ is known, the longitudinal stress σ_{zz} is determined directly from

$$\sigma_{zz} = \frac{\sigma E\left(u_{x,x} + u_{y,y}\right)}{(1+\sigma)(1-2\sigma)} = \sigma\left(\sigma_{xx} + \sigma_{yy}\right), \tag{5.89}$$

or

$$\sigma_{zz} = \sigma \nabla^2 \chi. \tag{5.90}$$

Much of elasticity theory is devoted to the subject of solving these various equilibrium equations under different boundary conditions (e.g. spherical and cylindrical, torsional and extensional) and material shapes (e.g. rods, plates, shells). It is useful to obtain experience solving a variety of such problems. We will continue here, however, by surveying other problems that are often overlooked in first exposures to continuum mechanics, but are of profound significance in applications: crystals, dislocations, and fracture. Other topics, such as stress corrosion and fatigue, are more the province of materials science, yet continuum properties play a fundamental role. The principal stress axis helps define the location where material failure is most likely to occur. This is often due to the presence of a catalytic agent such as minute quantities of water. The details of this process are highly geometry and history dependent.

5.7 Microscopic structure and dislocations

Crystalline structure is important in the earth sciences as well as engineering (since metals generally have a well-defined lattice character). This regularity on a microscopic level introduces a fundamental anisotropy into the problem and the free energy can no longer be assumed to be a quadratic form based solely on scalar

5.7 Microscopic structure and dislocations

constructs derived from the strain tensor. In general, we must assume a quadratic form

$$\mathcal{F} = \frac{1}{2} C_{ijkm}\, \varepsilon_{ij}\, \varepsilon_{km}, \qquad (5.91)$$

where the fourth-rank tensor C_{ijkm} which we introduced in (4.62) is now referred to as the *elastic modulus tensor* from which we derived the generalized Hooke's law

$$\sigma_{ij} = \frac{\partial \mathcal{F}}{\partial \varepsilon_{ij}} = C_{ijkm}\, \varepsilon_{km}. \qquad (5.92)$$

To proceed further, we must explicitly accommodate the specific geometry and symmetry properties of the underlying crystal structure of a specific mineral or metal, something which lies beyond the scope of this book. We conclude our discussion of crystalline structure and its influence upon continua by noting that the thermal expansion of isotropic materials which contributed to the strain according to

$$\varepsilon_{ij} = \frac{1}{3} \alpha\, (T - T_0)\, \delta_{ij} \qquad (5.93)$$

must now be replaced by the general (and anisotropic) contribution

$$\varepsilon_{ij} = \frac{1}{3} \alpha_{ij}\, (T - T_0). \qquad (5.94)$$

The second rank tensor α_{ij} is symmetric and, accordingly, expresses a preferred set of axes, i.e. the principal directions of α_{ij}, which relate macroscopic to microscopic behavior.

Dislocations are often thought of in the context of crystalline materials, but can sometimes be applied to many other materials. Basically, however, we are considering defects which influence the mechanical properties of crystals. In figure 5.2, we observe the role played by the insertion of an "extra" half-plane into the lattice. This so-called *wedge-dislocation* is one of the principal types of dislocation. We observe that the distortion in the lattice becomes very small as we go to substantial distance from the dislocation where the crystal planes fit in a relatively regular manner. Nevertheless, the deformation exists far from the dislocation. Any quantity, such as the displacement **u**, which is derived from an integral over a closed loop which passes through the dislocation, undergoes an irreversible change of one lattice period.

Another type of dislocation may be visualized as the result of "cutting" the lattice along a half-plane and then shifting the parts of the lattice on either side of the cut in opposite directions to a distance of one lattice period parallel to the edge of the cut, which is then called a *screw dislocation*. Such a dislocation converts the lattice planes into a helicoidal surface, like a spiral staircase without steps.

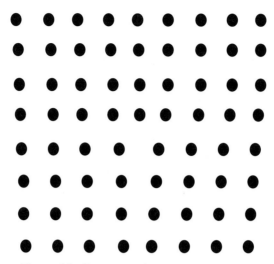

Figure 5.2 Geometry of a wedge dislocation.

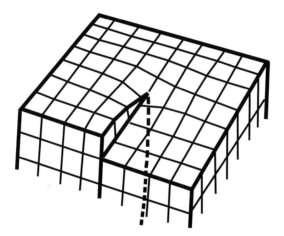

Figure 5.3 Geometry of a screw dislocation.

Macroscopically, a dislocation deformation of a crystal regarded as a continuous medium has the following general property: after a passage around any closed contour $\mathcal{L}$ which enclosed the dislocation line $\mathcal{D}$, the elastic displacement vector **u** receives a certain finite increment **b** which is equal to one of the lattice vectors in magnitude and direction. The constant vector **b** is called the *Burgers vector* of the dislocation and may be expressed as

$$\oint_{\mathcal{L}} du_i = \oint \frac{\partial u_i}{\partial x_k} dx_k = -b_i, \tag{5.95}$$

where the direction in which the contour is traversed and the chosen direction of the tangent vector to the dislocation line are assumed to be related by the "corkscrew rule." In reality, the dislocation line is a line of singularities of the deformation field – the strain tensor is no longer well-behaved but contains periodic singularities. Although these geometrical considerations were first developed in applications to ideal crystal structure, they are widely employed in describing geophysical environments. For example, the screw dislocation can be readily associated with a "strike-slip" fault.

The calculation of stress fields around dislocations and cracks is now tremendously complicated by the appearance of mathematical singularities. (Indeed, some of the linear approximations employed in the derivations are suspect.) Nevertheless, the harmonic or biharmonic character of the underlying continuum equations provides some insight into the nature of the stress field. In particular, at least in some situations, we can invoke the Green's functions for the appropriate harmonic operator – although general solutions (particularly in three dimensions) are not known. The stress in two dimensions, say for a penny-shaped crack, varies basically as $1/\sqrt{d}$, where d is the distance to the nearest crack tip with an amplitude which is described by the "stress intensity factor" as we shall see in chapter 9. Owing to the analytic difficulties hinted at above, fracture and crack mechanics remains a largely phenomenological and empirical science. In the last chapter, we will survey some of the issues confronting us in the description of fracture and friction, the fractal geometry of surfaces, and we will review the notion of scale invariance, and consider some of the observed earthquake scaling laws, including the Gutenberg–Richter and Omori laws, and some possible explanations. Now, we will turn our attention to classical fluid mechanics, geophysics fluid dynamics, and computational methods, providing a brief introduction to these topics.

Exercises

5.1 Explain why, at a fixed surface, we require that the displacement satisfies

$$\mathbf{u} = 0.$$

At a free surface with outward normal $\hat{\mathbf{n}}$, show why

$$\sigma_{ij}\hat{n}_i = 0.$$

5.2 Determine the deformation of a long rod of length ℓ standing on end vertically in a uniform gravitational field.

5.3 Suppose now that the rod is heated from below, e.g. it is standing on the coil of a stove, and that it (and the surrounding air) have a uniform temperature gradient oriented upward. Determine the deformation of the rod as in the previous problem.

5.4 Determine the deformation of a hollow sphere (with external and internal radii R_2 and R_1, respectively) with a pressure P_1 inside and P_2 outside. Hint: The displacement vector is radial.

5.5 Determine the deformation of a solid sphere (of radius R) in its own gravitational field. Hint: calculate the gravitational potential for a homogeneous sphere as a function of radius and, therefore, the pressure as a function of radius.

5.6 Determine the deformation of a sphere rotating uniformly about its z-axis.

5.7 Suppose that f designates one of the components of the displacement **u** for a monochromatic wave with wave vector **k**. We observed that f should satisfy the wave equation

$$\frac{\partial^2 f}{\partial t^2} = c^2 \nabla^2 f,$$

where c is the seismic velocity appropriate to the mode being considered, i.e. longitudinal or transverse, in a given medium. Suppose we are dealing with a material interface with different seismic velocities, say c_i, for $i = 1, 2$, designates the medium being considered. Across the interface, the frequency of a monochromatic wave and the amplitude of the tangential component of the wave vector are preserved. (The latter guarantees the conservation of energy.) Let θ and θ' be the angles of incidence and reflection (or refraction) and c, c' the velocities of the two waves. Show, for each case, that

$$\frac{\sin \theta}{\sin \theta'} = \frac{c}{c'},$$

thereby proving Snell's law for seismic waves.

5.8 Given the Lamé decomposition for Hooke's law

$$\sigma_{ij} = \lambda \delta_{ij} \varepsilon_{kk} + 2\mu \varepsilon_{ij},$$

compute the strain energy in terms of
(1) the components of ε_{ij}, and
(2) the components of σ_{ij}.

5.9 Consider a rod under homogeneous compression (say, P per unit area) along the z-axis. Evaluate ε_{11} and ε_{33} and, therefore, Young's modulus E.

5.10 Show for an isotropic elastic medium that

(1) $\dfrac{1}{1+\sigma} = \dfrac{2(\lambda+\mu)}{3\lambda+2\mu}$ and

(2) $\dfrac{\sigma}{1-\sigma} = \dfrac{\lambda}{\lambda + 2\mu}$,

where σ is Poisson's ratio.

5.11 A familiar theme in many science fiction movies is that an "NEO" or "near Earth object" is headed directly toward Earth. Suppose this asteroid has uniform density and a 10 km diameter. You plan to travel to the asteroid, drill a hole using a terawatt laser into its center where you will deposit a mid-size nuclear device in order to destroy it. You are assigned the task of calculating the solution to the associated continuum mechanical problem.

(1) You begin by noting that the displacement vector $\mathbf{u}$ will be strictly radial in spherical coordinates (r, θ, ϕ), with the origin at the center of the asteroid, so that

$$\mathbf{u} = u_r \mathbf{e}_r = \nabla \mathcal{U}\left(r^2, t\right),$$

where $\mathcal{U}$ is a function of the radius squared, and that u_r is a function only of the radius and time.

(2) Prove that the flow is strictly longitudinal.

(3) Prove that u_r satisfies

$$\frac{1}{c_\ell^2} \frac{\partial^2 u_r}{\partial t^2} = \nabla^2 u_r = \frac{1}{r^2} \frac{\partial}{\partial r}\left[r^2 \frac{\partial u_r}{\partial r}\right],$$

where we have used only the radial portion of the Laplacian.

(4) Suppose we assume that we can write

$$u_r(r, t) = \frac{U(r, t)}{r}.$$

Show that the associated differential equation for U has the form

$$\frac{1}{c_\ell^2} \frac{\partial^2 U(r, t)}{\partial t^2} = \frac{\partial^2 U(r, t)}{\partial r^2}.$$

(5) Show that $U(r, t)$ has a solution of the form $f(r - c_\ell t)$ where f can be *any* function of its argument, namely a radially outgoing traveling wave. Therefore, a solution to this problem can have the form

$$U(r, t) = \delta(r - c_\ell t),$$

where we have introduced the Dirac δ function which describes the explosive "seismic" pulse expanding through the asteroid.

6
Classical fluids

Fluid mechanics occupies an important niche in the study of the Earth. Fluid motions describe the behavior of the interior of this and other planets, as well as the motion of our respective oceans and atmospheres. Remarkably, fluid behavior can manifest some truly amazing properties. Van Dyke (1982) provides a visual compendium of these behaviors, describing the richness of flow patterns that can emerge. The study of fluid mechanics remains a venerable topic and there are a number of excellent textbooks available. Batchelor (1967) provides an authoritative introduction to the subject from the perspective of an applied mathematician, while Landau and Lifshitz (1987) does so from the viewpoint of theoretical physicists. Faber (1995) provides a modern treatment which is encyclopedic in scope but remains a relatively easy read. Fowler (2011) has published an encyclopedic volume addressing many flow problems encountered in geophysics.

Owing to their intrinsic nature, fluids can respond rather dramatically to subtle, almost imperceptible changes in their environment. We now appreciate that sensitivity to an initial set of conditions is the hallmark of *chaos*. Indeed, thermal convection is often cited as a source of chaos (Drazin, 1992) and the *Lorenz model*, a skeletal description of a fluid heated from below, as is the case in earth's atmosphere and mantle, has become the paradigm for chaotic behavior. The Lorenz model consists of three coupled ordinary differential equations (Strogatz, 1994; Drazin, 1992). Without elaboration, they are given by

$$\dot{x} = \sigma(y - x)$$
$$\dot{y} = rx - y - xz \qquad (6.1)$$
$$\dot{z} = xy - bz,$$

where x, y, and z are related to physical variables and emerge from a drastically simplified model of convection rolls in the atmosphere. Here, σ is related to the

Prandtl number and r is related to the *Rayleigh number*; we will discuss both of these dimensionless numbers later in this chapter, and b is a parameter related to the geometry of convection. A stability analysis of this system is outside the scope of this book, but a brief summary can provide some interesting insights. Despite the seemingly innocuous character of these equations, they can produce truly remarkable behavior and, for a certain physically relevant range of parameter values, exhibit chaos. An infinitesimal perturbation to the initial conditions results in a "trajectory" that departs exponentially rapidly from the original unperturbed trajectory. The implications of this sensitivity is profound: chaotic physical systems are intrinsically unpredictable in their nature. A chaotic system which consists of a large number of interacting components or possesses a large number of degrees of freedom, as is the case with the molecules in a fluid, also manifests what is called *complexity*. Another name for complexity is high degree of freedom chaos. The combination of chaos with a large number of degrees of freedom confers upon fluids a richness of character not readily apparent in other physical systems.

Two topics in fluid mechanics which emerge as a result are especially important. These include hydrodynamic stability and turbulence. Drazin (2002) provides an introduction to hydrodynamic stability while Drazin and Reid (2004) provide a more comprehensive treatment of the subject. Chandrasekhar (1961) treats not only hydrodynamic but hydromagnetic stability. Remarkably, there are many parallels – and important differences – between fluid flows and magnetohydrodynamic flows. The latter has a vital role in terms of the geodynamo and the generation of magnetic fields in the Earth's interior. An important outcome of many classes of hydrodynamic as well as hydromagnetic instability is *turbulence*. Tennekes and Lumley (1972) provide an excellent conceptual introduction to the subject, appealing to many simple yet compelling intuitive and theoretical arguments. Batchelor (1953) provides an authoritative treatment of the subject, blending deep mathematical insights with a strong physics-based intuition. Finally, Pope (2000) has developed an advanced textbook on the subject, embracing many topics rarely treated in other books. The narrative which now follows provides a transition from the continuum mechanical concepts already discussed to the behavior of fluids that is so critical to the Earth and related sciences.

6.1 Stokesian and Newtonian fluids: Navier–Stokes equations

The fundamental characteristic of a fluid, in contrast with a solid, is that the action of shear stresses, however small, will cause the fluid to deform continuously. At the end of this chapter we will offer some comments relating to turbulence and shock waves as well as to self-organization and solitary waves. The implication then is that the stress vector on an arbitrary element of surface at any point in an inviscid fluid is in the same direction as the normal $\hat{\mathbf{n}}$. Thus, we write

$$t_i^{(\hat{n})} = \sigma_{ij}\hat{n}_j = -P_0\hat{n}_i, \tag{6.2}$$

where the positive constant P_0 is the *hydrostatic pressure*. Accordingly, it follows that

$$\sigma_{ij} = -P_0 \delta_{ij}, \tag{6.3}$$

which indicates that for a fluid at rest the stress is compressive, that every direction is a principal stress direction at any point, and that the hydrostatic pressure is equal to the mean normal stress

$$P_0 = -\frac{1}{3}\sigma_{ii}. \tag{6.4}$$

The pressure is assumed related to the temperature T and density ρ by an equation of state having the form

$$F(P_0, \rho, T) = 0. \tag{6.5}$$

This describes how the kinetic energy on a microscopic scale, which is associated with the temperature and the density, is related to the rate of transfer of momentum in the fluid – the macroscopic meaning of pressure.

For a fluid in motion the shear stress is not usually zero and, in this case, we write

$$\sigma_{ij} = -P\delta_{ij} + \tau_{ij}, \tag{6.6}$$

where τ_{ij} is called the *viscous stress tensor*, which is a function of the motion and vanishes when the fluid is at rest. Here, we refer to the pressure P as the *thermodynamic pressure* and it is given by the same functional relationship with respect to T and ρ as that for the static pressure P_0 in the stationary state, namely

$$F(P, \rho, T) = 0. \tag{6.7}$$

Further, we observe that for a fluid in motion, the pressure is no longer the mean normal stress but is given by

$$P = -\frac{1}{3}(\sigma_{ii} - \tau_{ii}). \tag{6.8}$$

Since the viscous stress tensor vanishes for fluids at rest, it seems reasonable to assume that τ_{ij} is a function of the rate of deformation tensor D_{ij}. Expressing this symbolically, we write

$$\tau_{ij} = f_{ij}(\mathbf{D}); \tag{6.9}$$

6.1 Stokesian and Newtonian fluids: Navier–Stokes equations

this is called *Stokesian flow* in the general nonlinear case and is called *Newtonian flow* in the linear situation. The general linear Newtonian situation is analogous to the linear Hooke's law and can be represented by

$$\tau_{ij} = K_{ijkm}\, D_{km}, \tag{6.10}$$

where the fourth-rank tensor K_{ijkm} reflects the viscous properties of the fluid. Since fluids are experimentally confirmed to be isotropic, the 81 components of K_{ijkm} can be reduced to only two: a proportional dependence upon the velocity gradient tensor D_{ij}, as well as an isotropic dependence on its first invariant. Hence, we write

$$\sigma_{ij} = -P\,\delta_{ij} + \lambda^*\,\delta_{ij}\,D_{kk} + 2\mu^*\,D_{ij}, \tag{6.11}$$

where λ^* and μ^* are *viscosity coefficients*. This linear situation closely parallels that which we encountered in describing elastic media. From this equation, we see that the mean normal stress for a Newtonian fluid is

$$\frac{1}{3}\sigma_{ii} = -P + \frac{1}{3}\left(3\lambda^* + 2\mu^*\right)D_{ii} = -P + \kappa^* D_{ii}, \tag{6.12}$$

where

$$\kappa^* = \frac{1}{3}\left(3\lambda^* + 2\mu^*\right) \tag{6.13}$$

is known as the *coefficient of bulk viscosity*. The condition $\kappa^* = 0$ or, equivalently,

$$\lambda^* = -\frac{2}{3}\mu^* \tag{6.14}$$

is known as *Stokes' condition* and ensures that, for a Newtonian fluid at rest, i.e. $D_{ij} = 0$, the mean normal stress equals the (negative) pressure P.

In order to relate these definitions to our earlier stress-related nomenclature, we recall the deviator tensor S_{ij} according to

$$S_{ij} = \sigma_{ij} - \frac{1}{3}\delta_{ij}\,\sigma_{kk}, \tag{6.15}$$

for the stress, and

$$\beta_{ij} = D_{ij} - \frac{1}{3}\delta_{ij}\,D_{kk}, \tag{6.16}$$

for the rate of deformation. Accordingly, we obtain

$$S_{ij} + \frac{1}{3}\delta_{ij}\,\sigma_{kk} = -P\,\delta_{ij} + \frac{1}{3}\delta_{ij}\left(3\lambda^* + 2\mu^*\right)D_{kk} + 2\mu^*\,\beta_{ij}, \tag{6.17}$$

which can be conveniently split into the pair of constitutive equations

$$S_{ij} = 2\mu^*\,\beta_{ij}, \tag{6.18}$$

and
$$\sigma_{ii} = -3\left(P + \kappa^* D_{ii}\right). \tag{6.19}$$

The first of these equations relates the role of shear to the deviatoric stress, while the second associates the mean normal stress with the pressure and bulk viscosity which is often assumed to be zero.

We now wish to recall the basic equations of viscous flow in Eulerian and Lagrangian form:

- the continuity equation
$$\dot\rho + \rho\, v_{i,i} = \frac{\partial \rho}{\partial t} + v_i\, \rho_{,i} + \rho\, v_{i,i} = \frac{\partial \rho}{\partial t} + (\rho\, v_i)_{,i} = 0; \tag{6.20}$$

- the equations of motion
$$\sigma_{ij,j} + \rho\, b_i = \rho\, \dot v_i = \rho \left(\frac{\partial v_i}{\partial t} + v_j\, v_{i,j}\right); \tag{6.21}$$

- the constitutive equations
$$\sigma_{ij} = -P\, \delta_{ij} + \lambda^*\, \delta_{ij}\, D_{kk} + 2\,\mu^*\, D_{ij}; \tag{6.22}$$

- the energy equation
$$\rho\, \dot u = \rho \left(\frac{\partial u}{\partial t} + v_i\, u_{,i}\right) = \sigma_{ij}\, D_{ij} - q_{i,i} + \rho\, r; \tag{6.23}$$

- the kinetic equation of state
$$P = P(\rho, T); \tag{6.24}$$

- the caloric (or internal energy) equation of state
$$u = u(\rho, T); \quad \text{and} \tag{6.25}$$

- the heat conduction equation
$$q_i = -\kappa\, T_{,i}. \tag{6.26}$$

This system, together with the definition of the velocity deformation tensor, namely
$$2\, D_{ij} = v_{i,j} + v_{j,i}, \tag{6.27}$$

represents 22 equations in 22 unknowns: σ_{ij}, ρ, v_i, D_{ij}, u, q_i, P, and T. In situations where thermal effects are unimportant and a strictly mechanical problem is proposed, we employ only the first three of these equations, together with the definition for the velocity deformation tensor, and a temperature-independent form of the equation of state, which often emerges in isothermal or adiabatic environments,
$$P = P(\rho). \tag{6.28}$$

It is convenient to combine these field equations to produce a more compact formulation of viscous fluid problems. In particular, if we fold the constitutive and continuity equations into the equations of motion, we obtain

$$\rho \dot{v}_i = \rho \left(\frac{\partial v_i}{\partial t} + v_j \frac{\partial v_i}{\partial x_j} \right) = \rho b_i - P_{,i} + \left(\lambda^* + \mu^* \right) v_{j,ji} + \mu^* v_{i,jj}, \quad (6.29)$$

which are known as the *Navier–Stokes* equations for fluids. Note, despite the apparent linear character of these equations, that the inertial terms in the transport term $\dot{v}_i$, together with the potential for density variation, cause these equations to become highly nonlinear. If the Stokes' condition $\lambda^* = -\frac{2}{3} \mu^*$ is assumed, the Navier–Stokes equations reduce to the form

$$\rho \dot{v}_i = \rho \left(\frac{\partial v_i}{\partial t} + v_j \frac{\partial v_i}{\partial x_j} \right) = \rho b_i - P_{,i} + \frac{1}{3} \mu^* \left(v_{j,ji} + 3 v_{i,jj} \right). \quad (6.30)$$

If the kinetic equation of state is independent of temperature, i.e. $P = P(\rho)$, then the Navier–Stokes equations together with the continuity equation form a complete set of four equations in the four unknowns v_i and ρ.

In addition to satisfying the field equations above, it is essential that the solutions also satisfy the boundary and initial conditions on both traction and velocity components. The boundary conditions at a fixed surface require not only the normal, but also the tangential components of velocity to vanish because of the microscopic "boundary layer" that is established. This discussion, which has been restricted to smooth or *laminar flow*, requires further consideration to treat *turbulent flow* situations and shocks.

6.2 Some special fluids and flows

A number of situations arise in geophysical and engineering contexts which deserve particular attention. A *barotropic fluid* is one where the equation of state is independent of temperature. Two examples include *isothermal* fluids as well as *adiabatic* ones. In the latter case, the entropy is often a material property of the fluid and depends only on **X**, i.e. it is fixed for a given fluid element. Combining the second law with the equation of state produces a barotropic pressure dependence, often in the form of a power law $P \propto \rho^\gamma$. Another familiar situation emerges with *incompressible fluids*, where the fluid density ρ never changes. The continuity equation then reduces to $v_{i,i} = 0$, and the Navier–Stokes equations become

$$\rho \dot{v}_i = \rho \left(\frac{\partial v_i}{\partial t} + v_j \frac{\partial v_i}{\partial x_j} \right) = \rho b_i - P_{,i} + \mu^* v_{i,jj}. \quad (6.31)$$

These equations are frequently employed to describe flows in water, oil, and the Earth's mantle. *Inviscid* fluids cannot sustain shear stresses under any condition and both λ^* and μ^* vanish. Such *perfect fluids* satisfy the reduced Navier–Stokes equations

$$\rho \dot{v}_i = \rho b_i - P_{,i}, \tag{6.32}$$

often called the *Euler equations of motion*. Finally, an *ideal gas* obeys the ideal gas law

$$P = \rho R T, \tag{6.33}$$

where R is the "gas constant" for the particular gas under consideration. A more universal expression of this equation of state comes about when we write

$$P V = N k_B T, \tag{6.34}$$

where V defines the volume, N defines the number of independent fluid particles within that volume, k_B is Boltzmann's constant, and T is the temperature.

In contrast with these special fluids, let us consider some special kinds of flows. In the case of a *steady flow*, there is no explicit time dependence and the material (time) derivative of the velocity becomes

$$\dot{v}_i = v_j v_{i,j}. \tag{6.35}$$

Thus, for a steady flow, the Euler equations are modified to read

$$\rho v_j v_{i,j} = \rho b_i - P_{,i}. \tag{6.36}$$

Moreover, if the velocity field is zero everywhere, then the fluid is at rest and the situation is referred to as being *hydrostatic*. Assuming that a barotropic condition exists between ρ and P, we define a *pressure function* $\mathcal{P}$ according to

$$\mathcal{P}(P) \equiv \int_{P_0}^{P} \frac{dP'}{\rho}. \tag{6.37}$$

Now, if we may assume that the body forces are conservative, we may express them in terms of a potential function Φ by the relationship

$$b_i = \Phi_{,i}. \tag{6.38}$$

From the definition of the pressure function, it follows that

$$\mathcal{P}_{,i} = \frac{1}{\rho} P_{,i}, \tag{6.39}$$

so that the hydrostatic equilibrium equation now becomes

$$(\Phi + \mathcal{P})_{,i} = 0. \tag{6.40}$$

6.2 Some special fluids and flows

As a simple example, consider incompressible flow in a cylindrical pipe which is oriented along the z-axis and driven by a piston, so that the pressure gradient is uniform and in the z-direction – we assume that no other forces are acting. Thus, we write

$$\sigma_{ij} = -P(z)\,\delta_{ij} + 2\,\mu^*\,D_{ij}. \tag{6.41}$$

Furthermore, we assume that the flow is oriented strictly along the cylinder axis, although the velocity components themselves can vary according to their distance from the axis. Thus, the relevant equations for steady flow are simply

$$\rho v_j\, v_{i,j} = -P_{,i} + \mu^*\, v_{i,jj}. \tag{6.42}$$

The preceding symmetry considerations give us

$$v_z = v_z(x, y), \quad v_x = v_y = 0, \quad P_{,i} = -\alpha\,\delta_{i3}, \tag{6.43}$$

where α is a constant; we are implicitly assuming that the piston is driving the flow from left to right, the direction in which the pressure then drops. Hence, we simply get

$$v_{z,ii}(x, y) = \nabla^2 v_z(x, y) = -\frac{\alpha}{\mu^*}, \tag{6.44}$$

subject to the boundary conditions that $v_x = 0$ on the pipe walls. A more natural coordinate system here is cylindrical, and it is appropriate to assume azimuthal symmetry in the solution. Thus, $v_z(x, y) \to v_z(r, \theta) \to v_z(r)$ and we have

$$\frac{1}{r}\frac{d}{dr}\left[r\,\frac{dv_z(r)}{dr}\right] = -\frac{\alpha}{\mu^*}; \quad \text{with} \quad v_z(r = R) = 0, \tag{6.45}$$

where R is the cylinder radius. Since this equation is linear, we can integrate twice subject to the boundary conditions and obtain the solution

$$v_z(r) = \frac{\alpha}{4\,\mu^*}\cdot\left(R^2 - r^2\right) = \frac{\alpha}{4\,\mu^*}\cdot\left(R^2 - x^2 - y^2\right). \tag{6.46}$$

Thus, we observe that the velocity profile for this flow is parabolic. Further, it is easy to verify that this solution is incompressible, i.e. $\nabla \cdot \mathbf{v} = 0$.

Recall from equation (3.76) that the vorticity vector is $\boldsymbol{\omega}$ defined according to

$$\omega_i = \epsilon_{ijk}\,\partial_j\, v_k; \tag{6.47}$$

flows which are vorticity free are called *irrotational*. We can readily show that shear flows, on the other hand, have vorticity and are not irrotational. As a simple example, suppose $v_x(x, y, z) = \gamma\, y$, where γ is a constant, and the components $v_y(x, y, z)$ and $v_z(x, y, z)$ vanish. We observe that $\omega_z = -\gamma$ while $\omega_x = \omega_y = 0$, verifying our claim.

Suppose that the velocity field for a flow can be derived from a potential φ, which we call the *velocity potential*

$$\mathbf{v} = \nabla \varphi. \tag{6.48}$$

Given that $\nabla \times \nabla \varphi = 0$, we conclude that the velocity field for an irrotational flow can be derived from a potential. We refer to this situation as a *potential flow*. Irrotational flow becomes particularly important when considering the equations of motion for incompressible, inviscid flows, namely those defined by

$$\rho \left(\frac{\partial v_i}{\partial t} + v_j \frac{\partial v_i}{\partial x_j} \right) = \rho b_i - P_{,i}. \tag{6.49}$$

Recalling that incompressible flows are divergence free, $v_{j,j} = 0$, and assuming that the external body forces are derived from a potential, i.e. $b_i = \Phi_{,i}$, we now make use of the identity

$$\mathbf{v} \cdot \nabla \mathbf{v} = \nabla \frac{v^2}{2} - \mathbf{v} \times (\nabla \times \mathbf{v}) = \nabla \frac{v^2}{2} - \mathbf{v} \times \boldsymbol{\omega}. \tag{6.50}$$

Supposing that the flow is irrotational (i.e. $\boldsymbol{\omega} = 0$), we can combine the latter two equations to get

$$\rho \frac{\partial v_i}{\partial t} = -\frac{\partial}{\partial x_i} \left(\frac{\rho v^2}{2} + P + \rho \Phi \right), \tag{6.51}$$

where we have taken ρ to be constant. Thus, for steady flow, this shows that

$$\left(\frac{\rho v^2}{2} + P + \rho \Phi \right) = \text{constant}. \tag{6.52}$$

This result, known as *Bernoulli's principle*, can be employed to explain many seemingly complex, albeit irrotational, fluid phenomena. A familiar example is a vertically oriented pipe blowing air, e.g. a vacuum cleaner with the direction of flow reversed, supporting a ping-pong ball.

This remarkable combination of hydrostatic pressure and kinetic energy terms obtained in the previous expression provides a hint as to the microscopic nature of the pressure – could it not be associated with a microscopic equivalent? This can be elucidated by taking the general Navier–Stokes equations

$$\rho \left(\frac{\partial v_i}{\partial t} + v_j \frac{\partial v_i}{\partial x_j} \right) = \rho b_i - P_{,i} + (\lambda^* + \mu^*) \, v_{j,ji} + \mu^* \, v_{i,jj} \tag{6.53}$$

and adding it to v_i times the continuity equation, namely

$$v_i \frac{\partial \rho}{\partial t} + v_i \frac{\partial (\rho v_j)}{\partial x_j} = 0: \tag{6.54}$$

6.2 Some special fluids and flows

this yields

$$\frac{\partial (\rho v_i)}{\partial t} = -\frac{\partial}{\partial x_j}\left(\rho v_i v_j + P \delta_{ij}\right) + \rho b_i + \left(\lambda^* + \mu^*\right) v_{j,ji} + \mu^* v_{i,jj}. \quad (6.55)$$

Apart from the dissipative and the (external) body force terms, we observe the balance that emerges between $P \delta_{ij}$ and $\rho v_i v_j$. Using standard statistical mechanical averaging methods, we can show that the hydrostatic pressure is just $\langle \rho \, \delta v_i \, \delta v_j \rangle$ where δv_i describes the local microscopic *fluctuation* in the velocity field and where $\langle ... \rangle$ defines a local averaging over microscopic behavior. These considerations become particularly important when considering the phenomenology of turbulence.

Before discussing turbulent flows, it is useful to step back and consider the emergence of sound or acoustic waves. A particularly simple approach to the problem of sound waves is to consider an inviscid, barotropic fluid with no body forces acting upon it. Then, the equations of motion can be written

$$\rho \dot{\mathbf{v}} = -\nabla P(\rho) = -\frac{dP}{d\rho} \nabla \rho, \quad (6.56)$$

or

$$\frac{d}{dt}\mathbf{v} = -c^2(\rho) \nabla \ln(\rho), \quad (6.57)$$

where $c^2(\rho) \equiv dP(\rho)/d\rho$ defines what we will presently observe is the sound speed. In similar fashion, we can write the continuity equation as

$$\frac{d}{dt}\ln(\rho) = -\nabla \cdot \mathbf{v}. \quad (6.58)$$

Taking a second time derivative of the latter equation and making use of the equation (6.57) where we now ignore the inertial terms which are presumed to be small so that d/dt can be replaced by $\partial/\partial t$, we obtain the wave equation

$$\frac{d^2}{dt^2} \ln(\rho) = c^2(\rho) \nabla^2 \ln(\rho) + \text{higher-order terms}. \quad (6.59)$$

This result can also be obtained by using standard perturbation theory methods (Batchelor, 1967). The identification of $dP/d\rho$ with the sound speed squared is now complete.

Vortical behavior resides at the foundation of turbulence. Turbulence cannot occur without vortices. Consider, therefore, the *velocity circulation* Γ_C around a closed path in a fluid if it exists, namely

$$\Gamma_C = \oint v_i \, dx_i = \int_S \epsilon_{ijk} v_{k,j} \, \hat{n}_i \, d^2x. \quad (6.60)$$

We recognize the last term as the integral of the vorticity over the surface. Here we have employed Green's theorem in converting the line integral to an area integral over the surface S. This surface is bounded by a closed loop where $\hat{n}_i$ is a normal to that surface. Note that, for an irrotational fluid, the circulation vanishes. It follows directly that

$$\dot{\Gamma}_C = \oint \dot{v}_i \, dx_i + \oint v_i \, (d\dot{x}_i) = \oint \left(\frac{\partial v_i}{\partial t} + v_j \frac{\partial v_i}{\partial x_j} \right) dx_i + \oint v_i \, dv_i. \quad (6.61)$$

Observe that the second integral vanishes identically over a closed curve. For a barotropic, incompressible, inviscid fluid with conservative body forces, the quantity within parentheses in the first integral can be expressed as a gradient, i.e. as the derivative with respect to the ith coordinate. Hence, it too must vanish. This is *Kelvin's circulation theorem,* that Γ_C remains constant. Thus, a closed vortex behaves almost as though it were a fundamental entity.

The behavior of vortices can be explored more effectively by taking the curl of the Navier–Stokes equations (6.31) for an incompressible fluid. Since the curl of a gradient vanishes, a number of terms disappear and we are left with

$$\rho \left(\frac{\partial \omega_i}{\partial t} + v_j \frac{\partial \omega_i}{\partial x_j} + \omega_j \frac{\partial v_i}{\partial x_j} \right) = \mu^* \omega_{i,jj}. \quad (6.62)$$

This result is often called the *Taylor–Proudman theorem* (Pedlosky, 1979).

Suppose, for the moment, that we are considering an inviscid *two-dimensional flow*. Then, the $(\boldsymbol{\omega} \cdot \nabla) \mathbf{v}$ term and the right-hand side vanish, and our equation simply becomes

$$\dot{\omega}_i = 0. \quad (6.63)$$

Thus, vorticity is exactly preserved along particle trajectories (or "streamlines"). Thus, let us imagine our two-dimensional fluid being an ensemble of ideal point vortices, analogous to point particles. Since $\boldsymbol{\omega} = \nabla \times \mathbf{v}$ and $\nabla \cdot \mathbf{v} = 0$, it follows that $\mathbf{v}$ can be established by solving the equation $\nabla^2 \mathbf{v} = -\nabla \times \boldsymbol{\omega}$. This can be accomplished using Green's function methods (Arfken and Weber, 2005; Mathews and Walker, 1970). Thus, if we know the initial placement and strength of an ensemble of vortices, we can compute the associated velocity field and determine how each vortex moves from one instant of time to the next. Then, given the new positions of the vortices, a new velocity field can be established, and this process repeated to determine the velocity field at successive instants of time.

Suppose, now, we activate the viscosity term. Then, we observe that the right-hand side has the form $\mu^* \nabla^2 \boldsymbol{\omega}$ and describes the diffusive loss of vorticity. (Although the integrated vorticity is preserved, vortices with opposing spins are allowed to merge.) The point here is that we have introduced a dissipation mechanism, but observe that the $(\boldsymbol{\omega} \cdot \nabla) \mathbf{v}$ term cannot sustain the vorticity since the vorticity is perpendicular to the plane containing the flow. This result is an outcome of the *Taylor–Proudman theorem* and, briefly, shows that two-dimensional viscous flows are fundamentally dissipative and irreversible. The only way a flow can sustain itself is for the vorticity to couple with the velocity shear in this latter term, a phenomenon referred to as *vortex stretching*. This, in turn, can only happen in three dimensions and necessarily involves a complex velocity field whose shear can stimulate the growth of vorticity. Turbulence is fundamentally a three-dimensional phenomenon and balances the growth of vorticity with dissipation. Tennekes and Lumley (1972) provide an insightful conceptual review of turbulence and Batchelor (1953) develops the statistical theory while Pope (2000) presents a graduate-level text on the subject.

In real fluids, particularly on a planetary scale, dissipation rates, as reflected in effective viscosities, tend to be as much as seven orders of magnitude higher than what one might expect from molecular properties of the fluid. What is observed here is that vortical structures emerge wherein the intermediate or mesoscale behavior, i.e. smaller than the macroscopic scale, but larger than the microscopic molecular scale, shows statistical correlation. To better understand this, consider decomposing the velocity field into its macroscale and mesoscale components (Landau and Lifshitz, 1987; Tennekes and Lumley, 1972), that is

$$v_i \equiv V_i + \delta v_i, \tag{6.64}$$

where V_i represents a (locally) spatial average of v_i, and δv_i is the difference between the local flow speed and its local average. Earlier, we referred to the δv_i as "fluctuations" as they describe the complicated behavior of small units of the fluid, in contrast with microphysical effects which are on a molecular level.

Consider, now, the adaptation of the Navier–Stokes equations (6.55) and the term

$$-\frac{\partial}{\partial x_j}\left(\rho\, v_i\, v_j\right) \to -\frac{\partial}{\partial x_j}\left(\rho\, V_i\, V_j + \rho\, V_i\, \delta v_j + \rho\, \delta v_i\, V_j + \rho\, \delta v_i\, \delta v_j\right). \tag{6.65}$$

We now redefine the averaging operation $\langle ... \rangle$ to be over intermediate length scales – in contrast with the microscopic scales used before to distinguish molecular from macroscopic behavior. If we take such an average over the above terms, we will observe that terms of the form $V_i \delta v_j$ will necessarily vanish as V_i is essentially constant while δv_j is fluctuating. Thus, the only contributions to survive from the above include

$$-\frac{\partial}{\partial x_j}\left(\rho\, V_i\, V_j + \rho\, \langle \delta v_i\, \delta v_j \rangle\right). \tag{6.66}$$

This illustrates that shorter-scale fluctuations (both meso and microscopic) can contribute to the bulk evolution of the flow, and that there are average effects which can survive from these fluctuations leaving a nontrivial imprint on the fluid's behavior. Empirically, the smaller-scale behavior corresponds to vortical structures which, in turn, argues for the δv_i and δv_j terms to be *correlated*.

In particular, it is observed using scaling arguments based on the notion that typical short-scale velocities are a substantial fraction of the sound speed, while the corresponding characteristic distance scales are eddy sizes. Taken together, we approximate this behavior by assuming that this correlation is driven by the velocity shear in the mean flow, that is

$$\langle \delta v_i \, \delta v_j \rangle \approx -\eta \frac{\partial V_i}{\partial x_j}, \tag{6.67}$$

where the empirically estimated constant η has the dimensionality of a diffusion or viscosity coefficient (i.e. length^2/time) and is often called an "eddy viscosity." Inserting this into the former expression, we now observe that the forcing which emerges from the inertial terms can be approximated by

$$-\frac{\partial \left(\rho \, V_i \, V_j \right)}{\partial x_j} + \rho \, \eta \, \frac{\partial^2 V_i}{\partial x_j^2}, \tag{6.68}$$

giving rise to a new diffusion term in the empirically modified Navier–Stokes equations. The process where we have attempted to separate the behavior into different size scales is sometimes referred to as developing *moment equations*; however, this invariably results in more variables than equations to solve them, a situation referred to as the *closure problem* (Tennekes and Lumley, 1972). This problem is generally accommodated by introducing an empirical relation, such as the one we have invoked, to close the system of equations.

We have already hinted at one essential facet of turbulence, that it is an inherently random process where probabilistic or statistical considerations are important. Hence, we use averages to separate the *mean flow* from the smaller-scale behavior. By the same token, we have observed that vortices occupy a pivotal role yet, despite their inherent directionality, must become anisotropic in some global sense. An alternate approach to the moment based methodology discussed above attempts to amalgamate the statistical flavor of this problem with an isotropic characterization within the context of the many scales contained within the problem. Indeed, what is seen in many turbulence problems is a *cascade* where large-scale eddies, stirred up by the largest scale motions available to a system, break down into smaller eddies transporting their energy to these smaller structures in the process, whereupon these smaller eddies also undergo breakdown carrying energy until one reaches molecular size scales where further breakdown is no longer possible and the energy carried down is lost as heat. We refer to this as a *direct cascade*. There are also situations, notably in planetary atmospheres, where the cascade may proceed from the smallest scales to the largest; this is referred to as an *inverse cascade*. We refer to the largest scale sizes as the "production subrange" and the smallest scale sizes as the "dissipation subrange," while the cascade over intermediate scale sizes is referred to as the "inertial subrange." In order for this process to maintain some kind of time stationarity, it is essential that the process look the same on all scales, i.e. it must be *self-similar*, and that the rate of transport of energy must be scale-independent.

Consider, now, an eddy with characteristic size λ and speed (of rotation) v_λ. We employ a "dissipation rate" at this scale denoted by ϵ_λ which we presume is independent of λ and represents the energy (per unit mass) per unit time that passes or cascades through that size scale. Dimensionally, we associate a quantity of order v_λ^2 with the energy per unit mass available to a given eddy. The characteristic time scale for this eddy (say, for rotation) is of order λ/v_λ – hence, the dissipation rate at a given size scale λ satisfies

$$\epsilon_\lambda \approx \frac{v_\lambda^3}{\lambda}. \tag{6.69}$$

Assuming now that the dissipation rate is scale-independent, i.e. that $\epsilon_\lambda = \epsilon$, we write

$$v_\lambda \approx (\epsilon \lambda)^{1/3}, \tag{6.70}$$

an approximation widely known as Obukhov's law. The assumption of scale invariance is present in the underlying equations and in most observations. We relate the wavelength to the wavenumber k according to $k = 2\pi/\lambda$, whereupon it follows that

$$v_k \approx \left(\frac{\epsilon}{k}\right)^{1/3}. \tag{6.71}$$

The power spectrum $S(k)$, or power spectral *density* in N dimensions, is a measure of the energy at a given wavenumber. The spectral power therefore is v_k^2, divided by k^N to reflect the dimensionality, and satisfies

$$S(k) \approx \frac{v_k^2}{k^N} \approx \epsilon^{2/3} k^{-N-2/3}, \tag{6.72}$$

which is generally known as the "Kolmogorov–Obukhov law" for isotropic turbulence. Many complex flows in three dimensions have this power-law index $-11/3$. Other flows, characterized for example by an angular momentum or vorticity cascade, satisfy rather different power laws.

To conclude this chapter, we present an analytic example of a problem that has enjoyed widespread use as a paradigm both for turbulence and for shock waves. Burgers' equation (Whitham, 1974) is a simplified one-dimensional version of the incompressible Navier–Stokes equations discussed earlier. Owing to its low dimensionality, it cannot manifest vorticity but it is, nevertheless, highly nonlinear. Burgers' equation is commonly written as

$$\frac{dv(x,t)}{dt} \equiv \frac{\partial v(x,t)}{\partial t} + v(x,t) \cdot \frac{\partial v(x,t)}{\partial x} = \nu \frac{\partial^2 v(x,t)}{\partial x^2}, \qquad (6.73)$$

where $u(x,t)$ is the velocity and $\nu > 0$ is equivalent to a coefficient of kinematic viscosity and where the initial condition

$$u(x,0) = u_0(x) \qquad (6.74)$$

is specified on some domain, e.g. the infinite interval. Ignoring for the moment the diffusion term on the right-hand side of equation (6.73), this has the appearance of a momentum conservation principle. From the Lagrangian form of this expression, we see that each "particle" in this flow preserves its velocity. However, we note in situations where $\partial v/\partial x < 0$ that these particles will overtake each other and a shock will emerge as the particles collide. The introduction of the diffusive term in principle prevents this from happening.

In the full fluid equations described earlier, shocks do form and discontinuities in the flow properties emerge, so-called *shock fronts*. The treatment of shocks, beyond what we are about to present using Burgers' equation, resides outside the scope of this book but Landau and Lifshitz (1987) and Faber (1995) provide excellent treatments. The fundamental ingredient in the analysis of shocks is the application of the full fluid equations expressed in "conservative form." For example in section (4.2), we derived the continuity equation (4.25) which we express in one-dimensional form

$$\frac{\partial \rho}{\partial t} + \frac{\partial}{\partial x}(\rho v) = 0. \qquad (6.75)$$

Integrating over all space, the conservation of mass is guaranteed. Suppose that a shock wave is present propagating with a speed v_s. Since the fluid equations are valid in any reference frame, let us re-express the continuity equation in the frame of the shock, i.e. in a reference frame moving at speed of v_s. In that reference frame, where x' and v' designate the transformed position and velocity, we observe that

$$\frac{\partial \rho}{\partial t} + \frac{\partial}{\partial x'}(\rho v') = 0. \qquad (6.76)$$

It follows, therefore, on either side of the shock, that the flux of mass $\rho v'$ must be the same; otherwise, the discontinuity would produce a singularity in the equation. Using subscripts "l" and "r" to designate the left- and right-hand sides of the shocks, it follows that

$$\rho_l v'_l = \rho_r v'_r. \qquad (6.77)$$

Similar results can be obtained from the equations of motion, the energy equations, the constitutive equation, the equation of state, etc. In many important cases such as an adiabatic fluid, these *jump conditions* or *Rankine–Hugoniot conditions* yield a specific relationship between the different physical quantities on either side of a shock. Detailed calculations are provided elsewhere. We return to our discussion of Burgers' equation.

Hopf (1950) and Cole (1951) independently obtained closed-form solutions to equation (6.73), and a detailed discussion is provided in Whitham (1974). Since this equation is dissipative, we expect and can easily show that the energy in the flow $\int v^2(x,t)\, dx$ is steadily decreasing. Suppose for the moment that the velocity can be described by a velocity potential φ according to

$$v(x,t) \equiv \frac{\partial \varphi(x,t)}{\partial x}, \tag{6.78}$$

whereupon we find that Burgers' equation has become transformed into

$$\frac{\partial^2 \varphi(x,t)}{\partial x\, \partial t} + \frac{\partial \varphi(x,t)}{\partial x} \cdot \frac{\partial^2 \varphi(x,t)}{\partial x^2} = \nu \frac{\partial^3 \varphi(x,t)}{\partial x^3}. \tag{6.79}$$

We observe that the middle term can be expressed as the derivative of a product thereby allowing us to express the first integral of this expression as

$$\frac{\partial \varphi}{\partial t} + \frac{1}{2}\left(\frac{\partial \varphi}{\partial x}\right)^2 = \nu \frac{\partial^2 \varphi}{\partial x^2}. \tag{6.80}$$

We note that we can combine both spatial derivative terms into one if we make the substitution

$$\varphi \equiv -2\nu \ln \phi, \tag{6.81}$$

which yields the *diffusion equation*

$$\frac{\partial \phi}{\partial t} = \nu \frac{\partial^2 \phi}{\partial x^2}. \tag{6.82}$$

Remarkably, we have now converted this paradigm for nonlinear fluid flow into a linear partial differential equation. Importantly, the general solution for the diffusion equation on the infinite interval is well-known and treated in standard textbooks (Arfken and Weber, 2005; Mathews and Walker, 1970) using Green's function methods. By converting our initial conditions for Burgers' equation using equations (6.78) and (6.81) into initial conditions for $\phi(x,0)$ by integrating twice, we can write (Mathews and Walker, 1970)

$$\phi(x,t) = \int_{-\infty}^{\infty} \frac{1}{\sqrt{4\pi \nu t}} \exp\left[-\frac{(x-x')^2}{4\nu t}\right] \phi(x',0)\, dx' \tag{6.83}$$

and $v(x, t)$ can be calculated for all time t using equations (6.78) and (6.81). We can readily verify that (6.83) is a solution of the diffusion equation (6.82) and, in the limit $t \to 0$, recovers the initial data. This is the *Green's function* solution to this problem.

In our discussion of the Kolmogorov–Obukhov relation for fully developed, isotropic turbulence, we introduced an energy source on the largest-size scales – analogous to those forces that drive convection in the interior of the Earth as well as in its atmosphere and oceans. The emergent energy cascade from the production range through the inertial range and down to the dissipation range is the hallmark of turbulence. Since Burgers' equation does not possess a simple extension to higher dimension, it does not provide us with further insight into direct cascades. However, the explicit and nonlinear role of diffusion upon the flow results in the energy resident at shorter length scales being transferred to larger scales as well as being dissipated. In this way, Burgers' equation has been shown by Newman (2000) to provide insight into the *inverse cascade*.

As a final discussion point in our treatment of fluid mechanics, we wish to turn to the notion of collective modes of behavior and self-organization, i.e. situations in which the different length scales resident in a fluid flow act in a cooperative way to produce well-defined structures that resist disruption. These have come to be known as *solitary waves* and some very special ones, known as *solitons*, have some truly remarkable conservation properties. We have discussed briefly the influence of diffusion, a fundamentally dissipative process. There exists a complementary process known as *dispersion* which acts on different length scales in a very different way. Dispersion causes the propagation speed of different length scales to vary. The nature of Burgers' equation can be fundamentally altered if the diffusive term $\nu \, \partial^2 v / \partial x^2$ is replaced by $-\partial^3 v / \partial x^3$. Historically, a slightly different representation for the advective $v \, \partial v / \partial x$ term was applied, and we introduce the (normalized) *Korteweg–de Vries* or *KdV* equation

$$\frac{\partial v(x,t)}{\partial t} + 6 v(x,t) \frac{\partial v(x,t)}{\partial x} + \frac{\partial^3 v(x,t)}{\partial x^3} = 0. \quad (6.84)$$

(The factor of 6 can be eliminated if we redefine v by $v/6$ revealing the parallel with Burgers' equation.) Drazin and Johnson (1989) provide an excellent introduction to the Korteweg–de Vries equation and Ablowitz and Segur (1981) provide an exhaustive survey of the subject of solitary waves, solitons, and related topics.

Solitary waves describe phenomena that maintain their structure as they travel. Evidently, the study of solitary waves originated with observations made by J. Scott Russell while riding on horseback in 1834 beside a narrow barge channel (Ablowitz and Segur, 1981). He noticed, using Russell's words, "a large solitary elevation, a rounded, smooth and well defined heap of water, which continued its course

along the channel apparently without change of form or diminution of speed...."
At that time, no mathematical theory emerged yielding the equation underlying
Russell's observations. More than half a century elapsed before Korteweg and de
Vries conclusively established (6.84) as the defining equation for the process. The
KdV equation became a topic of intense investigation in the 1960s and thereafter,
with the demonstration that it possessed a family of solitary wave solutions

$$v(x,t) = 2k^2 \text{sech}^2\left[k\left(x - 4k^2 t - x_0\right)\right], \tag{6.85}$$

where k is a free parameter that defines the reciprocal of the length scale of the
soliton as well as its speed, $4k^2$, while x_0 defines the center of the distribution
when $t = 0$.

Another remarkable discovery emerged for the KdV equation from the observation that two different solitary waves, initially widely separated, could pass through each other unscathed, albeit with their respective values of x_0 altered. The latter observation prompted many researchers to look for possible conservation laws in equation (6.84). For example, by inspection, we note that the integral $\int v(x,t)\,dx$ is preserved. What was truly remarkable, however, was the discovery of an infinite number of conservation laws for solitons. Solitons are solitary waves that possess an infinite number of conservation laws. The history underlying these discoveries is chronicled in Drazin and Johnson (1989) and Ablowitz and Segur (1981).

It is very unlikely that solitons with many of the conservation properties of the KdV equation are generic and occur often in nature. However, there is a substantial body of phenomenological evidence that solitary waves and other collective modes of nonlinear behavior are commonplace in nature and that characteristic shapes remain robust and can be preserved in a wide array of natural phenomena. We will now return from our general considerations of fluid mechanics to its application to the Earth sciences.

Exercises

6.1 Write down the equations for one-dimensional motion of an ideal fluid in terms of the variables **X** and t where X is the reference coordinate and describes a *Lagrangian variable* and defines the X co-ordinate of a fluid particle at $t = 0$. In particular, show that

$$\rho\left(\frac{\partial x}{\partial X}\right)_t = \rho_0(X) \tag{6.86}$$

where ρ_0 defines the initial or reference density distribution. Further, show that Euler's equation becomes

$$\left(\frac{\partial v}{\partial t}\right)_X = -\frac{1}{\rho_0}\left(\frac{\partial P}{\partial X}\right)_t. \tag{6.87}$$

These equations are widely used in one-dimensional numerical simulations.

6.2 Consider the velocity potential

$$\Phi = \frac{x_2 x_3}{r^2}, \tag{6.88}$$

where $r^2 = x_1^2 + x_2^2$ (i.e. we are using cylindrical coordinates). Show that this satisfies the Laplace equation $\nabla^2 \Phi = 0$. Derive the velocity field and show that this flow is both incompressible and irrotational.

6.3 Consider a velocity field defined by

$$\mathbf{v} = \left(x_3 - x_2^2\right)\hat{\mathbf{e}}_2 + (x_3 + x_2)\hat{\mathbf{e}}_3. \tag{6.89}$$

Show, for *any* closed contour in the x_2-x_3 plane, that the circulation vanishes.

6.4 Consider a situation where we have an undisturbed incompressible, inviscid flow with no applied hydrostatic pressure or body forces such that the undisturbed velocity field $\mathbf{v}(\mathbf{x})$ is simply $\hat{\mathbf{z}} v_0$. Suppose we now introduce a hard, sphere of radius R into the flow—for convenience, we will assume that the center of the sphere is at the origin. We wish to derive the resulting steady, irrotational flow. In particular, we will assume (owing to the lack of viscosity) that the velocity component normal to the sphere's surface vanishes. Begin by showing that the velocity field can be described by a velocity potential Φ, i.e.

$$\mathbf{v}(\mathbf{x}) = \nabla \Phi, \tag{6.90}$$

subject to the boundary conditions that

$$\lim_{\mathbf{x} \to \infty} \nabla \Phi = v_0 \hat{\mathbf{z}}, \tag{6.91}$$

and, momentarily resorting to spherical coordinates,

$$\left(\frac{\partial \Phi}{\partial r}\right)_{r=R} = 0. \tag{6.92}$$

[Hint. It is sufficient to consider axisymmetric flow, i.e. $\Phi(\mathbf{x}) = \Phi(r, \theta)$.] Show that Φ can be expressed via

$$\Phi(r, \theta) = \sum_{n=0}^{\infty} \left(a_n r^n + b_n r^{-(n+1)}\right) P_n(\cos \theta), \tag{6.93}$$

Exercises

where the $P_n(\zeta)$, for $-1 \leq \zeta \leq 1$, define the geophysically practical Legendre polynomials. Show that the second of our boundary conditions cause all b_n coefficients to be expressible in terms of the a_n coefficients according to

$$b_n = a_n \frac{n}{n+1} R^{2n+1}. \tag{6.94}$$

Then, show that the first of our boundary conditions cause all a_n to vanish apart from a_1 and that, from matching the boundary conditions at great distance,

$$a_1 = v_0, \tag{6.95}$$

so that the final velocity potential is given by

$$\Phi(r, \theta) = v_0 r \left[1 - \frac{1}{2}\left(\frac{R}{r}\right)^3 \right]. \tag{6.96}$$

One can employ the modified velocity field to determine the stresses placed on the sphere to estimate the effective drag that a "Stokes sphere" (such as a meteor entering our atmosphere) encounters.

6.5 A viscous fluid occupies the region between two long concentric cylinders of radius R_1 and R_2 ($R_1 < R_2$) aligned along the z-axis. If the cylinders rotate with different angular velocities ω_1 and ω_2, find the steady-state velocity field, which is known as *Couette flow* and is highly relevant to considerations of the relative motion of the earth's core and mantle.

6.6 We showed that a viscous incompressible fluid with vorticity

$$\boldsymbol{\omega} \equiv \nabla \times \mathbf{v} \tag{6.97}$$

obeys the equation

$$\frac{d\boldsymbol{\omega}}{dt} \equiv \frac{\partial \boldsymbol{\omega}}{\partial t} + (\mathbf{v} \cdot \nabla) \boldsymbol{\omega} = (\boldsymbol{\omega} \cdot \nabla) \mathbf{v} + \mu^* \nabla^2 \boldsymbol{\omega}. \tag{6.98}$$

Consider an axisymmetric flow (in cylindrical geometry where the coordinates are r, ϕ, and z)

$$\mathbf{v}(\mathbf{x}, t) = V(r, t) \hat{\boldsymbol{\phi}}. \tag{6.99}$$

In particular, show that the vorticity $\boldsymbol{\omega}$ satisfies

$$\boldsymbol{\omega}(r, t) = \mu^* \frac{1}{r} \frac{\partial (rV)}{\partial r}, \tag{6.100}$$

and that

$$\frac{\partial \boldsymbol{\omega}}{\partial t} = \mu^* \left(\frac{\partial^2 \boldsymbol{\omega}}{\partial r^2} + \frac{1}{r} \frac{\partial \boldsymbol{\omega}}{\partial r} \right). \tag{6.101}$$

Then, consider the special initial configuration

$$\omega(r, 0) = \omega_0 \delta(\mathbf{x}) = \omega_0 \delta(x)\delta(y). \tag{6.102}$$

Show that the time-dependent solution is

$$\omega(r, t) = \frac{\omega_0}{4\pi \mu^* t} \exp\left(-\frac{x^2 + y^2}{4\mu^* t}\right). \tag{6.103}$$

[Hint. If you are not familiar with the Green's function for the diffusion equation, show that the latter satisfies (4.47) and the initial conditions (which also requires that you show that the vorticity integrated over all space is consistent with the initial conditions).] Finally, show that the circulation of the flow is *not* preserved and explain why it decays.

6.7 Suppose that our flow corresponds to solid-body rotation, i.e.

$$\mathbf{v}(\mathbf{x}, t) = \mathbf{\Omega} \times \mathbf{x} \tag{6.104}$$

where $\mathbf{\Omega}$ is a constant vector which describes the rotation rate of the object. Show for this case that

$$\boldsymbol{\omega} = 2\mathbf{\Omega} \tag{6.105}$$

Moreover, show that the divergence of the velocity field is zero.

6.8 Suppose that the density of a fluid undergoing flow is a constant, neither changing in time nor in position. Show, using the continuity equation, that

$$\nabla \cdot \mathbf{v}(\mathbf{x}, t) = 0 \tag{6.106}$$

for "incompressible flow."

6.9 In the Euler force equation, we encounter the term

$$\{\mathbf{v}(\mathbf{x}, t) \cdot \nabla\} \mathbf{v}(\mathbf{x}, t).$$

Using the Levi-Civita symbol, show that

$$(\mathbf{v} \cdot \nabla) \mathbf{v} = \nabla \left(\frac{1}{2} v^2\right) - \mathbf{v} \times \boldsymbol{\omega} \tag{6.107}$$

using the vorticity $\boldsymbol{\omega}$ defined earlier.

6.10 Suppose that we are dealing with an incompressible fluid that is subject to viscous forces, but gravity and pressure play no important role. The flow velocity $\mathbf{v}(\mathbf{x}, t)$ satisfies the equation

$$\frac{\partial \mathbf{v}(\mathbf{x}, t)}{\partial t} + \{\mathbf{v}(\mathbf{x}, t) \cdot \nabla\} \mathbf{v}(\mathbf{x}, t) = \eta \nabla^2 \mathbf{v}(\mathbf{x}, t), \tag{6.108}$$

where η is the diffusion rate for velocity. Take the curl of (6.108) and show that

$$\frac{\partial \boldsymbol{\omega}(\mathbf{x},t)}{\partial t} + \{\mathbf{v}(\mathbf{x},t) \cdot \nabla\} \boldsymbol{\omega}(\mathbf{x},t) = \{\boldsymbol{\omega}(\mathbf{x},t) \cdot \nabla\} \mathbf{v}(\mathbf{x},t) + \eta \nabla^2 \boldsymbol{\omega}(\mathbf{x},t). \tag{6.109}$$

The first term on the right hand side of the equation is referred to as a "vortex-stretching" term and is fundamental to turbulence theory; unless it provides a significant transfer of energy among different length scales, so-called fully-developed isotropic turbulent behavior cannot happen.

6.11 Suppose we multiply equation (6.109) by $\boldsymbol{\omega}(\mathbf{x},t)$ and integrate over all space. Show that the enstrophy

$$\mathcal{E} \equiv \frac{1}{2} \int_V |\boldsymbol{\omega}|^2 \, d^3x \tag{6.110}$$

must necessarily decrease over time, so long as the vortex-stretching term in $\boldsymbol{\omega} \cdot \nabla \mathbf{v}$ does not intervene, which it cannot (why?) for two-dimensional flows. This is the essence of the Taylor-Proudman theorem.

7
Geophysical fluid dynamics

In the previous chapter, we focused on the behavior of fluids in an inertially stationary environment. The Earth, as well as other planets and stars, on the other hand, rotate at a relatively constant rate thereby introducing a time scale and a length scale, namely the length of day and the Earth's radius, respectively, into the problem. Moreover, rotation introduces non-inertial forces, as we saw in section 1.6, that interact with other forces that are present, particularly gravity. This is further complicated by issues ranging from heat flow, variations in density, the role of viscosity, and the presence of topographically complex boundaries. Pedlosky (1979) provides an exhaustive survey of the field of geophysical fluid dynamics (GFD). Many arenas of investigation have emerged in response to these theoretical advances. Atmospheric and oceanographic flow, owing to their practicality, have preserved a prominent place in contemporary science. Houghton (2002) focuses on the physics of atmospheres while Holton (2004) presents a more meteorologically-based perspective. Marshall and Plumb (2008) explore the combined roles of atmosphere and ocean and their interaction, exploring also their long-term contributions to climate dynamics. Gill (1982) focuses on the atmosphere–ocean dynamics. Given that Earth's hydrosphere has almost three orders of magnitude more mass than its atmosphere, these interaction effects can be profound. Ghil and Childress (1987) investigate a set of topics emerging from atmospheric dynamics, dynamo theory, and climate dynamics. Finally, Schubert *et al.* (2001) explore the role of mantle convection in the Earth and planets, while Turcotte and Schubert (2002) synthesize fluid and solid behaviors into the field of geodynamics. Fowler (2011) has recently published a monumental treatment of many problems that emerge not only in geophysical fluid dynamics but in many other Earth science contexts.

7.1 Dimensional analysis and dimensionless form

The subject of geophysical fluid dynamics is based on the principles of continuum mechanics that we have already discussed, where we now introduce a new

7.1 Dimensional analysis and dimensionless form

ingredient of fundamental importance: the Earth, as well as the other planets, is a rotating platform and the governing dynamical equations must be modified to accommodate this feature. This added ingredient provides an essential complication. However, some simple techniques, based on "dimensional analysis," have evolved for evaluating the importance of this effect.

All physical quantities contain the attributes (in different amounts) of mass, length and time. It is useful to employ square brackets "[" and "]" to designate the three types according to powers of M, L, and T where the latter three symbols designate mass, length, and time, respectively. For example, if c is the speed of sound, e.g. of transverse waves as discussed in the previous chapter, then

$$[c] = \frac{L}{T}. \tag{7.1}$$

More complicated examples follow in a natural way. For example, the dimensionality of the kinetic energy of a particle of mass m and velocity v is given by

$$\left[\frac{1}{2}mv^2\right] = M^1 L^2 T^{-2}. \tag{7.2}$$

Any energy quantity has this particular dimensionality Thus, even quantities such as the charge e of an electron have a well-defined dimensionality – since the energy of two electrons separated by a distance r is e^2/r, it follows then that

$$[e] = M^{1/2} L^{3/2} T^{-1}. \tag{7.3}$$

Indeed, upon reflection, essentially all physical quantities have a characteristic dimensionality.

Sometimes, it is convenient to decompose the dimensionality of a quantity according to the powers of M, L, and T involved, say α, β and γ. Thus, we can in general write

$$[\ldots] = M^\alpha L^\beta T^\gamma, \tag{7.4}$$

where ... represents some physical quantity. Interestingly, while many of the quantities in different areas of physics have seemingly unusual dimensionalities, that is not the case in continuum mechanics. Strains, being a measure of a displacement length relative to some benchmark length, have no dimensionality. Stresses and pressures have the same dimension as a force per unit area and, therefore,

$$[\sigma] = M L^{-1} T^{-2}. \tag{7.5}$$

As an outcome of this, the Lamé coefficients and bulk moduli possess the same dimensionality as stresses. The dimensionality of other geophysical and continuum mechanical quantities are readily calculable.

A special, but very important, situation emerges if $\alpha = \beta = \gamma = 0$, which we refer to as being *dimensionless*. Dimensionless numbers occupy an important role in many areas of physics, including geophysical fluid dynamics. As an example derived from modern physics, the quantity

$$\frac{e^2}{\hbar c} \approx \frac{1}{137.04} \qquad (7.6)$$

is known as the "fine structure constant," where $\hbar$ is Planck's constant and c is the speed of light. The fine structure constant describes the ratio of the "speed" of the most tightly bound electron orbiting a hydrogen atom *relative* to the speed of light. Since this quantity is so small and relativistic effects are proportional to this quantity squared, i.e. are of order 10^{-4}, special relativity is of minor importance in quantum mechanics and in radiative transfer theory. The gravitational equivalent of this quantity, namely $Gm_p^2/\hbar c$ where G is the universal gravitational constant and m_p is the mass of a proton, can be used in planetary science applications. In particular, it allows us to establish that the interior of a planet, even one like Jupiter where the pressure can exceed 1 megabar, is fundamentally different from that of a star.

In fluid dynamics, one of the most fundamental dimensionless quantities is the *Reynolds number Re* which defines the relative role of the inertial terms in the incompressible Navier–Stokes equations to the viscous dissipation,

$$Re = \left[\frac{\rho \mathbf{v} \cdot \nabla \mathbf{v}}{\mu \nabla^2 \mathbf{v}}\right] = \left[\frac{D v \rho}{\mu}\right]. \qquad (7.7)$$

Here, the viscosity μ when divided by the density ρ has a dimensionality equivalent to that of a diffusivity L^2/T – it is often called the *kinematic viscosity* and designated by the symbol ν. We have employed the quantity D to designate a characteristic length, while v designates a characteristic velocity. For example, in the case of simple pipe flow, D could be the pipe diameter and v could be the flow speed along the central axis. Observe, as well, that

$$[D v] = L^2 T^{-1}. \qquad (7.8)$$

Physically, Re denotes the relative importance of inertial forces to viscous ones. Put another way, we might think of

$$\left[\frac{\mu}{\rho D}\right] = \left[\frac{\nu}{D}\right] = \frac{L}{T} \qquad (7.9)$$

as a natural velocity associated with the viscous drag. If the true fluid velocity is much greater than this, we can expect that the shear in the flow is sufficiently great that the character of the flow is fundamentally altered. This, as we have seen before, is an essential feature of turbulence. Moreover, depending on the fluid, a

7.1 Dimensional analysis and dimensionless form

Reynolds number in excess of 1000–2000 is almost certainly a guarantee of turbulent behavior.

In the geophysical context, two dimensional quantities of fundamental importance include the Earth's radius

$$R \approx 6.4 \times 10^8 \text{cm}, \tag{7.10}$$

and the rotation rate of the Earth

$$\Omega = \frac{2\pi}{86\,400} \approx 7.3 \times 10^{-5} \text{s}^{-1}. \tag{7.11}$$

In the next section, we will introduce the Rossby and Ekman numbers.

A final issue that we wish to pursue in this section is that of rendering equations in non-dimensional form. Consider, for example, the incompressible version of the Navier–Stokes equations

$$\frac{\partial \mathbf{v}}{\partial t} + \mathbf{v} \cdot \nabla \mathbf{v} = \nu \nabla^2 \mathbf{v}. \tag{7.12}$$

Since mass is not explicitly involved in this equation, i.e. $\alpha = 0$ for all terms present, it is sufficient to consider a characteristic length D and time τ for the problem. Here, D and τ should be natural length and time scales, e.g. the diameter of the pipe and the time it takes, say, for a "typical" molecule to travel that distance. Then, we render each quantity dimensionless by defining its dimensionless equivalent, denoted by primed quantities. For example, we replace the role of $\mathbf{v}$ by

$$\mathbf{v}' = \frac{\mathbf{v}\tau}{D}. \tag{7.13}$$

Similarly, the diffusivity which has a characteristic L^2/T dimensionality can be written

$$\nu' = \frac{\nu\tau}{D^2}. \tag{7.14}$$

Note that $1/\nu' = Re$, the dimensionless Reynolds' number. Finally, the gradient operator ∇ (as well as its related divergence and Laplacian operators) can be replaced using

$$\nabla' = D\,\nabla. \tag{7.15}$$

Pulling everything together and using $t' = t/\tau$, we then obtain

$$\frac{\partial \mathbf{v}'}{\partial t'} + \mathbf{v}' \cdot \nabla' \mathbf{v}' = \nu' \nabla'^2 \mathbf{v}'. \tag{7.16}$$

Dropping the primes for convenience, we observe that this dimensionless equation has precisely the same form as the original dimensional form, but we note that the diffusion coefficient is now dimensionless. It is simply the reciprocal of the

Reynolds number. Now, we turn our attention to geophysical flows and the role of rotation.

7.2 Dimensionless numbers

Given the foregoing considerations, let us write in material derivative form the relevant force equation for a fluid observed in an inertial, i.e. non-rotating, reference frame where both gravity and viscosity are present:

$$\rho \frac{d\mathbf{v}}{dt} = -\nabla P + \rho \mathbf{g} + \mu \nabla^2 \mathbf{v}, \qquad (7.17)$$

where P is the atmospheric pressure. The stress tensor is assumed to be isotropic and hydrostatic $\sigma_{ij} = -P \delta_{ij}$. The local gravitational acceleration is given by $\mathbf{g}$. Since we are generally concerned with the left-hand side expressed in a rotating coordinate system, i.e. that of an observer co-rotating with his home planet, we should remind ourselves of the meaning of the Helmholtz theorem. In particular, when we considered the Lagrangian description of a continuum $\mathbf{x}(\mathbf{X}, t)$, we identified the velocity

$$\mathbf{v} \equiv \frac{d\mathbf{x}(\mathbf{X}, t)}{dt} \qquad (7.18)$$

with a combination of rotation (of coordinates or of the body) and deformation. In particular, the effect of rotation could be expressed

$$\left. \frac{d\mathbf{x}(\mathbf{X}, t)}{dt} \right|_{\text{rot}} = \mathbf{\Omega} \times \mathbf{x}(\mathbf{X}, t). \qquad (7.19)$$

Here, we have elected to use the vector symbol $\mathbf{\Omega}$ to represent the rotation rate. Quantitatively, $\Omega = 2\pi/86{,}400 \text{ sec} \approx 7.3 \times 10^{-5} \text{ s}^{-1}$ for the Earth. Consequently, we can now distinguish between the time derivative of a vector quantity as seen from space, i.e. inertial frame, and that as seen from the Earth. Therefore, we have

$$\left. \frac{d}{dt} \right|_{\text{inertial}} = \left. \frac{d}{dt} \right|_{\text{earth}} + \mathbf{\Omega} \times, \qquad (7.20)$$

where we view the latter expression in "operator form." Thus, the inertial velocity that we need in the force equation is the inertial velocity

$$\mathbf{v}_{\text{inertial}} = \mathbf{v}_{\text{earth}} + \mathbf{\Omega} \times \mathbf{x}_{\text{earth}}, \qquad (7.21)$$

where $\mathbf{x}_{\text{earth}}$ denotes the particle's position on the *rotating* Earth. Similarly,

$$\left. \frac{d\mathbf{v}_{\text{inertial}}}{dt} \right|_{\text{inertial}} = \left. \frac{d\mathbf{v}_{\text{inertial}}}{dt} \right|_{\text{earth}} + \mathbf{\Omega} \times \mathbf{v}_{\text{inertial}}. \qquad (7.22)$$

Finally, we need to obtain the left-hand side of this equation strictly in terms of quantities that we as observers measure from the surface of the Earth. Hence, it follows that

$$\left.\frac{d\mathbf{v}_{\text{inertial}}}{dt}\right|_{\text{inertial}} = \left.\frac{d\mathbf{v}_{\text{earth}}}{dt}\right|_{\text{earth}} + 2\mathbf{\Omega} \times \mathbf{v}_{\text{earth}} + \mathbf{\Omega} \times (\mathbf{\Omega} \times \mathbf{x}_{\text{earth}}). \tag{7.23}$$

Importantly, we recognize this expression as being essentially the same as equation (1.98) obtained by largely geometrical methods.

Given we understand that all quantities are measured relative to our rotating viewpoint, we will now drop the "earth" subscript and refer to the acceleration term solely as $\frac{d\mathbf{v}}{dt} + 2\mathbf{\Omega} \times \mathbf{v} + \mathbf{\Omega} \times (\mathbf{\Omega} \times \mathbf{x})$. The first of these terms is natural while the second term, known as the *Coriolis force*, reflects the influence of evaluating the force in a non-inertial frame. Two contributions emerge, one each from the time derivatives of position and of velocity in the Coriolis force. This Coriolis term is of profound importance in geophysical applications. The third term, which can be written

$$\mathbf{\Omega} \times (\mathbf{\Omega} \times \mathbf{x}) = \mathbf{\Omega}(\mathbf{\Omega} \cdot \mathbf{x}) - \Omega^2 \mathbf{x}, \tag{7.24}$$

has a simple and direct dependence upon spatial position. Hence, it is often expressed in the form of a potential which is added, in turn, to the gravitational potential responsible for the $\rho\,\mathbf{g}$ term – together, the gravitational potential and rotational term are known as the *geopotential*. The rotational part of the potential has the form of a simple harmonic oscillator in the plane perpendicular to the rotation axis (as seen above) where the "spring frequency" is just the rotation rate of the Earth. It is convenient to express the total force due to gravity and rotation in the form

$$\mathbf{g}' = \mathbf{g} - \mathbf{\Omega}(\mathbf{\Omega} \cdot \mathbf{x}) + \Omega^2 \mathbf{x}; \tag{7.25}$$

the prime is usually dropped and $\mathbf{g}$ usually refers to the combination. For most applications, $\mathbf{g}$ may be regarded to be in the local vertical direction. Hence, our fundamental equation (7.17) can be written

$$\rho\frac{d\mathbf{v}}{dt} + 2\rho\,\mathbf{\Omega} \times \mathbf{v} = -\nabla P + \rho\,\mathbf{g} + \mu\,\nabla^2 \mathbf{v}. \tag{7.26}$$

It is particularly convenient to separate this equation into its vertical and horizontal components. This explicitly segregates the role that gravity has in the description of the atmosphere, oceans, or mantle. (In the latter situation, buoyancy effects related to density stratification are most important and the equations undergo substantial simplification if consideration of gravitational influences is confined to buoyancy, the so-called *Boussinesq approximation*.) In the horizontal plane, assuming negligible inertial and frictional effects ($d\mathbf{v}/dt \approx 0 \approx \mu \nabla^2 \mathbf{v}$), there emerges an approximate force balance

$$2\rho\, \mathbf{\Omega} \times \mathbf{v} = -\nabla P; \qquad (7.27)$$

we refer to the velocity $\mathbf{v}$ calculated from this expression as the *geostrophic* approximation (Holton, 2004). With this component-based separation of terms, we must further distinguish between those terms that are essentially independent of time, i.e. "static," and those which depend explicitly on time, i.e. "dynamic."

Inertial terms emerging from $d\mathbf{v}/dt$, namely $\mathbf{v} \cdot \nabla \mathbf{v}$, may in many cases be ignored. To establish this, consider the relative importance of the inertial and Coriolis terms, namely $v/2\Omega D$, where we again employ D and v to mean characteristic distance and velocity scales. This dimensionless number, sometimes denoted ε, is given by

$$\varepsilon = \frac{v}{2\,\Omega\, D}, \qquad (7.28)$$

and is called the *Rossby number* and signifies the transition in behavior from ordinary inertial flow to flow dominated by planetary rotation via the denominator term Ω. Low Rossby numbers indicate that rotational effects are important. It follows, further, that such effects are particularly important when D is very large, i.e. of a planetary scale. An important example of this dynamical, cf. static, effect is the so-called "planetary" or Rossby waves which travel around planets like the Earth or Venus in four days. It is this flow that gives Venus its peculiar upper atmospheric character and produces disturbances in polar regions in the Earth's upper atmosphere. Another dynamical effect balances viscous drag with Coriolis forces. The ratio of viscous to Coriolis forces scales as

$$E = \frac{\nu}{2\,\Omega\, D^2}, \qquad (7.29)$$

where $\nu = \mu/\rho_0$, and E is known as the *Ekman number*. Viscous effects are especially important in regions where shear can have an important role over small scales, namely near the bottom of an ocean or atmosphere where topography can interact strongly with the flow. Pedlosky (1979) and others treat the subject of dynamical effects, including Rossby waves and the Ekman layer.

Before moving on, we should think again about the significance of the three dimensionless numbers that we have introduced for describing geophysical fluid dynamics, the Reynolds, Rossby, and Ekman numbers. In particular, we must remember that we are dealing with fluid *flows*, the behavior of large numbers, often many times Avogadro's number, of particles. On the other hand, one or more dimensionless numbers, albeit useful in characterizing a flow, do not present the whole story – after all, there are essentially an infinite number of degrees of freedom present in such systems (Tennekes and Lumley, 1972). Hence, it is important to recognize the utility of dimensionless numbers without overstating their importance. They are particularly useful in describing the emergence of qualitative change or transition in behavior, but obviously cannot describe everything about a flow, which is the product of the behavior of myriad numbers of molecules. We return now to the topic of statics in atmospheres and oceans.

In order to pursue an investigation of static effects, it is essential that we consider the order of magnitude of each of the terms in the governing equations. We can ignore quantitatively small terms in developing an understanding of these processes. As before, we replace the role of ∇ by D^{-1} where D is the relevant characteristic distance scale. Further, we employ typical velocities and representative Eddy viscosities – molecular viscosities as we have seen earlier do not adequately describe the complex behavior of real fluids which are *organized* in the form of vortices or eddies. The separation is most easily accomplished in the vertical direction where we find that the dominant terms are the pressure gradient $-\nabla P$ and the (geo)potential force $\rho \mathbf{g}$. Hence, we write the *hydrostatic equilibrium equation*

$$\frac{dP}{dz} = -\rho g. \tag{7.30}$$

Here, we employ z to denote the local vertical; x denotes the west-to-east direction while y denotes the south-to-north direction. Recalling the ideal gas law

$$P = \frac{\rho k_B T}{m}, \tag{7.31}$$

where k_B is Boltzmann's constant, T is the temperature, and m is the mean molecular mass of the fluid, it follows that

$$-\frac{1}{P}\frac{dP}{dz} = \frac{mg}{k_B T} \equiv \frac{1}{H}. \tag{7.32}$$

We note that the quantity H has the dimensionality of length and represents the e-folding length for the pressure. H is generally called the *pressure scale height*. When combined with the thermal speed, the scale height provides an effective frequency of oscillation, known as the *Brunt–Vaisala frequency* (Holton, 2004; Houghton, 2002; Pedlosky, 1979).

The density scale height is the same if temperature is uniform, i.e. isothermal. When the temperature is allowed to vary, e.g. in adiabatic environments, the density scale height will be smaller, e.g. by a factor of γ for adiabatic situations where γ is the specific heat ratio. The temperature gradient, in an adiabatic environment, can be shown to be constant, and is often referred to as the "lapse rate." It can be shown that a fluid which is strongly heated from below is driven into convective motion and the *Rayleigh–Taylor instability*, a result of the fact that a dense fluid cannot be supported on top of a less-dense one, will cause the fluid's temperature structure or "profile" to become adiabatic.

In the horizontal direction, the essential balance is that between the Coriolis force and the pressure gradient. This is often expressed via the "geostrophic approximation"

$$2\mathbf{\Omega} \times \mathbf{v} = -\frac{1}{\rho} \nabla P. \tag{7.33}$$

Since we are now concerned only with the upward component of Ω, this equation is often written

$$\mathbf{v} = \frac{f}{\rho} \hat{\mathbf{k}} \times \nabla P, \tag{7.34}$$

where $\hat{\mathbf{k}}$ is a unit vector in the z-direction, and f is the "Coriolis parameter" that emerges from the geometry of the previous equation, namely

$$f = 2\Omega \sin\phi, \tag{7.35}$$

where ϕ is the latitude. Flows are said to be "quasi-geostrophic" if they satisfy in an approximate way this expression. An immediate feature of this approximation is that local pressure maxima and minima are surrounded by circulating flows. For example in the Northern hemisphere low pressure regions are associated with counter-clockwise motion. (We note that f changes sign in going from one hemisphere to the other.) The association of low pressure regions with cyclones, hurricanes, etc., gives rise to these flows being called *cyclonic*; anti-cyclonic flows persist around high-pressure regions. In situations where D, the characteristic size scale is small, such as in a tornado, the inertial terms must also be taken into account. Quasi-geostrophic motion is the fundamental feature of atmospheric and oceanic flows and is the basis of many other books (Holton, 2004; Houghton, 2002; Marshall and Plumb, 2008; Vallis, 2006; Gill, 1982; Ghil and Childress, 1987).

Before concluding this chapter, two other issues deserve mention. We observed earlier that dimensionless numbers can describe the transition of one kind of behavior to another. Convection is associated with heating a fluid from below. When thermally induced forces prevail, i.e. those that make hot fluids rise and exceed resisting dissipative ones, then convective "rolls" can form. This ratio of forces, which lies beyond the scope of the present treatment, is known as the *Rayleigh number*; patterns of convective rolls are associated with "Rayleigh–Bénard convection" (Vallis, 2006; Gill, 1982; Ghil and Childress, 1987). To properly explore this issue, we must return to our earlier considerations of heat flux and its associated rate and the balance between diffusion of different molecular constituents and of heat. The dimensionless ratio of these two diffusive processes, molecular and thermal diffusion, is known as the *Prandtl number*. Such issues are of profound

Exercises

importance in the ocean where salt, which is present in proportions of 30-40 units per 1000, has a very important role, and the competition between the diffusion of heat and the diffusion of salt water can produce "double diffusive instabilities" that produces "salt fingers." Double diffusive convection occurs as well in the Earth's mantle (Schubert *et al.*, 2001).

The second issue emerges from dynamics. The essential methodology here is that we take the fundamental equations, calculate the static balance both in the horizontal and in the vertical directions, and then linearize the time-dependent equations around the static solutions. This isolates the wave motion and frequency as a function of the wavelength of the "disturbance," although the wavelength itself is not always important. For example, simple buoyancy in the atmosphere is characterized by the sound speed c_s, the mode by which perturbations propagate, and the pressure scale height H; the ratio of these two quantities defines the *Brunt–Vaisala frequency*, which is the frequency at which the atmosphere bobs up and down. These and other advanced topics are the subjects of other books.

Exercises

7.1 In our discussion of dimensional analysis, we have discussed the need to introduce order-of-magnitude estimates of physical quantities. Discuss the applicability of each of the following quantities in the context of the different dimensionless numbers that we have introduced.

1. $H \approx 10^4$ m, the vertical scale for the atmosphere (scale height) and oceans
2. $L \approx 10^3$ km, the horizontal scale for continents and oceans
3. $U \approx 10$ m s^1, the horizontal velocity scale
4. $W \approx 1$ cm s^{-1}, the vertical velocity scale
5. $v \approx 300$ m s^{-1}, the thermal/sound speed
6. $T = L/U \approx 10^5$ s, the (daily) time scale

Consider their application in calculation of the Reynolds number, Rossby number, Ekman number, and Prandtl number. (Be prepared to look up estimates for viscosity, topographical roughness, etc. as required.)

7.2 The dimensional numbers that appear in the governing partial differential equations often regulate the flow. For example, the diffusion constant v has the dimensions $[v] = L^2/T$. Turning that scaling relationship inside-out, it is interesting to consider the possibility that the relevant length scale ℓ for a diffusing quantity at a given time scale as

$$\ell \approx \sqrt{vt}. \qquad (7.36)$$

Equation (6.82) for the diffusion of a quantity ϕ in one-dimension is a case in point,

$$\frac{\partial \phi}{\partial t} = \nu \frac{\partial^2 \phi}{\partial x^2};$$

the Gaussian inside the integral appearing in equation (6.83), which we presented without proof, reveals this scaling property. We will now demonstrate the scaling in (7.36).

1. We shall assume that the integral $\int_{-\infty}^{\infty} \phi(x,t)\, dx$ is finite, and that no significant flow of the diffusing quantity extends to infinity. For convenience, we will think of this latter integral as the "mass" M of the system as a function of times. Prove that M is a constant.
2. Then, show that its "first moment" or mean

$$\mu(t) \equiv \frac{1}{M} \int_{-\infty}^{\infty} x \phi(x,t)\, dx \qquad (7.37)$$

is also a constant for the distribution.
3. Then, show that the "second moment" or variance for the system

$$\sigma^2(t) \equiv \frac{1}{M} \int_{-\infty}^{\infty} (x-\mu)^2 \phi(x,t)\, dx \qquad (7.38)$$

satisfies

$$\frac{d\sigma^2(t)}{dt} = \nu \qquad (7.39)$$

so that

$$\sigma^2(t) = 2\nu t + \sigma^2(0). \qquad (7.40)$$

From (7.38), it follows that σ^2 corresponds to mean-squared length $\approx \ell^2$ of the system, thereby verifying that our simple scaling relationship (7.36) applies.

7.3 We can now apply our scaling ideas to more complicated situations. Consider, for example, the impact of a small meteorite on the surface of a terrestrial planet and the propagation of the ensuing shock wave into the surrounding atmosphere. (This problem is analogous to the "point-blast explosion" of classical fluid dynamics.) We shall assume that the energy transferred by the blast to the atmosphere $\mathcal{E}$; this excludes seismic wave energy radiated away, and other non-adiabatic losses. Moreover, we shall assume that the mass of the meteorite and any vaporized surface material is negligible compared with the mass swept up by the expanding hemispherical shock in the enveloping atmosphere which has mass density ρ. (The atmospheric blast is approximately a hemisphere; the lower half of the blast vaporizes and seismically activates the rick below.) Finally, we shall assume that the energy transferred to the shocked gas is much greater than its original thermal content. We observe that the flow is characterized by the dimensional quantity $\mathcal{E}/\rho$ which has the dimensions

$$[\mathcal{E}/\rho] = L^5/T^2. \tag{7.41}$$

Accordingly, provide a justification for the scaling law

$$R_s(t) \approx \left(\frac{\mathcal{E}}{\rho}\right)^{1/5} t^{2/5} \tag{7.42}$$

where R_s is the shock radius of the emergent blast wave. It is now known that the exact solution to the fluid dynamical equations preserve this scaling and that the relevant density, velocity, and temperature profiles are simply functions of $r/R_s(t)$ multiplied by appropriate powers of time t. Such universal solutions which maintain their appearance over all time are known as *self-similar* solutions.

7.4 Consider the equation (7.26) for the Coriolis force in the absence of external forces, i.e. where the pressure and geopotential gradients vanish. Let us explore the dynamics of a particle moving in this environment perpendicular to the angular velocity $\mathbf{\Omega}$ of the system. This is sometimes referred, for reasons that will become evident, as the "inertial circles problem."

1. Show by ignoring the pressure, gravitation, and viscosity terms that the governing equation has the form

$$\frac{d\mathbf{v}}{dt} + 2\mathbf{\Omega} \times \mathbf{v} = 0. \tag{7.43}$$

2. Accordingly, by taking the dot product of the latter equation with $\mathbf{v}$ show that v^2 is a constant.

3. We now write

$$\mathbf{v} = \frac{d\mathbf{R}}{dt} \tag{7.44}$$

where $\mathbf{R}$ is in the plane perpendicular to $\mathbf{\Omega}$. Therefore, show that

$$\frac{d}{dt}\left(\frac{d\mathbf{R}}{dt} + 2\mathbf{\Omega} \times \mathbf{R}\right) = 0. \tag{7.45}$$

which has as its solution

$$\frac{d\mathbf{R}}{dt} + 2\mathbf{\Omega} \times \mathbf{R} = \mathbf{c}, \tag{7.46}$$

where $\mathbf{c}$ is a (vector) constant of integration that is perpendicular to $\mathbf{\Omega}$.

4. Then, show that our problem reduces to finding the solution to

$$\frac{d\mathbf{r}}{dt} + 2\mathbf{\Omega} \times \mathbf{r} = 0 \qquad (7.47)$$

where

$$\mathbf{R} = \mathbf{r} - \frac{1}{2\Omega^2}\mathbf{\Omega} \times \mathbf{c}. \qquad (7.48)$$

5. Finally, take the dot product of $\mathbf{r}$ with equation (7.47) to show that r^2 is a constant and, thereby, describes circular motion with angular frequency 2Ω. The constant cross product term in the last equation corresponds to the center of this gyro-motion.

8
Computation in continuum mechanics

Most of this book, in its pursuit of studying continuum mechanics, focused on linearized problems and their decomposition into relatively simple superpositions of solutions. Real-world problems, in contrast, are much more complicated. The calculation of exact solutions to linearized problems in the face of complicated geometries very often can become computationally intensive. Further, realistic problems often introduce substantial nonlinearity – this is especially true in applications involving fluids – rendering such calculations inaccurate, if not incorrect. Moreover, the application of computational methods, however, is not trivial. Computer arithmetic, unlike computer algebra such as that performed by Mathematica and Maple, is executed with a finite number of digits of accuracy (Higham, 2002). Typically, there are 16 digits in double precision arithmetic in programming languages such as C++ and FORTRAN 95, but as great as 25 in MATLAB. Continuum mechanical problems involving matrices, particularly those of large rank, can lose many digits of accuracy due to *ill-conditioned* matrices.[1] Adding to the arithmetic limitations of computers, the more common tasks of solving nonlinear ordinary and, especially, partial differential equations of continuum mechanics present a formidable issue. The operative differential equations were formulated in the mathematical limit of certain differential quantities going to zero, i.e. infinitesimal quantities that emerge in the evaluation of derivatives. Computers, on the other hand, only work in the realm of the finite quantities. Effective computation, therefore, required the development of a broad array of approximate methodologies to address the fundamental conceptual difference emerging from the finite vs the infinitesimal, yet do so efficiently in light of the arithmetic limitations posed by computers. Computation, arguably, is an art as well as a science.

Fröberg (1985) provides an illuminating survey of many standard mathematical issues that also emerge in continuum mechanics as well as an introduction

---

[1] For example, the fitting of a twelfth-degree polynomial to observational data introduces what is called a Hilbert matrix and all 16 digits of accuracy will be lost.

to the numerical methods that can help address them. Kincaid and Cheney (2009) provide a more rigorous and advanced analysis of numerical techniques. The texts by Gear (1971) and by Richtmyer and Morton (1967) are regarded as a classic in providing the conceptual framework for ordinary and partial differential equations, respectively, and their numerical integration, focusing as well on the issues of accuracy pertaining to numerical methods. The book by Peyret and Taylor (1990) provides an important survey of computational methods for fluid flow, while the handbook by Peyret (2000) is also helpful. Finally, as the mastery of computational methods for use in continuum mechanics also requires becoming familiar with the implementation of methods on computers, the book by Butt (2007) provides an overview of many numerical methods and their implementation in MATLAB. Learning a high-level computer language such as C++ or FORTRAN 95 involves a major investment of time before the user can efficiently develop working codes whose results are verifiable and whose errors can be estimated. For this reason, I routinely recommend that students learn MATLAB first, gain some experience in solving related computational problems and then, if necessary, graduate to a high-level programming language. Now, we turn to the focus of this chapter, computation in continuum mechanics.

8.1 Review of partial differential equations

The problems of continuum mechanics are essentially representable by partial differential equations, differential equations with two or more independent variables. Most partial differential equations are of three basic types: elliptic, hyperbolic, and parabolic (Fox, 1962).

Elliptic equations are often called potential equations and result from problems where the unknown potential might be temperature, electrical potential, or a similar quantity. Elliptic equations are also the steady solutions of diffusion equations, and they require boundary values in order to determine the solution. Often, elliptic equations are linear and have Green's function solutions. The latter often can be expressed strictly in terms of the function (and its normal derivative) on the boundary. This is the basis for the *boundary element method*.

Hyperbolic equations are sometimes called wave equations, since they often describe the propagation of waves. They require initial conditions, where the waves originate, as well as boundary conditions to describe how the wave and boundary interact; for instance, the wave might be reflected or absorbed. These equations can be solved, in principle, by the *method of characteristics* – although not always practical, characteristics provide some important insights into the nature of the problem. Parabolic equations are often called diffusion equations since they describe the diffusion of some quantity. The dependent variable usually represents the density of

8.1 Review of partial differential equations

the substance. These equations require initial conditions (what the initial concentration of the substance is) as well as boundary conditions (to specify, for example, whether the substance can cross the boundary or not).

For equations of order higher than two, i.e. where derivatives of order three or more appear, the above classification is not adequate. However, most higher-order equations in continuum mechanics can be re-expressed in terms of second-order equations of derived variables. For example, the biharmonic equation $\Delta\Delta u = 0$ can be written $\Delta u = f$ and $\Delta f = 0$. The method of characteristics can also be used to simplify such higher-order equations.

The method of characteristics emerges from considering quasi-linear partial differential equations, i.e. one or more partial differential equations linear in the first derivatives of the dependent variables, with no higher-order derivatives present. For example, the (dimensionless) wave equation

$$\frac{\partial^2 u}{\partial t^2} = \frac{\partial^2 u}{\partial x^2}, \qquad (8.1)$$

where the propagation velocity is assumed to be one, can be decomposed into

$$\frac{\partial u}{\partial t} = \frac{\partial v}{\partial x} \quad \text{and} \quad \frac{\partial v}{\partial t} = \frac{\partial u}{\partial x}, \qquad (8.2)$$

which has the necessary quasi-linear form. The required initial data, $u(x, 0) = \mathcal{F}(x)$ and $u_t(x, 0) = \mathcal{G}(x)$ can also be expressed in terms of u and v; the boundary conditions can similarly be adapted. Note that geometrical insights can also simplify the problem. For example, if we use new coordinates ζ and ξ defined by

$$\zeta = x + y \quad \text{and} \quad \xi = x - y, \qquad (8.3)$$

our wave equation can be written

$$\frac{\partial^2 u}{\partial \zeta \, \partial \xi} = 0, \qquad (8.4)$$

whose complete solution can always be expressed as

$$u(\zeta, \xi) = f(\zeta) + g(\xi), \qquad (8.5)$$

or

$$u(x, t) = f(x + t) + g(x - t). \qquad (8.6)$$

Here, the functions f and g are arbitrary and can be related directly to the initial conditions. The former indicates that the "directions" $x - t$ and $x + t$ are in some sense special and are, in fact, characteristic lines for the solution. Independent information is carried along each of the characteristics.

The notion of characteristics is often ignored in modern treatments of numerical methods, an unfortunate oversight since they can provide important analytic understanding into the nature of complex problems. As a simple example of a characteristic, consider the quasi-linear equation

$$a(x, y, u)\, u_x(x, y) + b(x, y, u)\, u_y(x, y) = c(x, y, u). \tag{8.7}$$

The variable y is general; often, it may be regarded to be the time variable t. By inspection, i.e. consideration of the exact differential $u_x\, dx + u_y\, dy = du$, it follows that the following situation is particularly important:

$$\frac{dx}{a} = \frac{dy}{b} = \frac{du}{c}. \tag{8.8}$$

The first half of this, namely

$$\frac{dx}{dy} = \frac{a(x, y, u)}{b(x, y, u)} \tag{8.9}$$

may be regarded as the differential equation which defines the characteristic line, while

$$\frac{du}{dy} = \frac{c(x, y, u)}{b(x, y, u)} \tag{8.10}$$

defines the evolution of u along that line. If we think of y as being time, this defines the evolution of u in time on a characteristic line, which has the form of a "velocity" $dx/dt = a/b$. This coupled pair of ordinary differential equations provides a complete description of that characteristic. The above was expressed in a manner appropriate for one space dimension plus time, although the general idea can be extended to more than one space dimension. Similarly, the above was expressed in one dependent variable, the *scalar* problem, but can be extended to several dependent variables, the *vector* problem.

As a practical example of characteristics for continuum systems, consider an *isentropic* fluid in one dimension where $P = P(\rho)$, and where we define

$$c^2(\rho) \equiv \frac{dP(\rho)}{d\rho}. \tag{8.11}$$

We take the continuity and inviscid force equations

$$\frac{\partial \rho}{\partial t} + \frac{\partial}{\partial x}(\rho v) = 0 \quad \text{and} \quad \rho\left(\frac{\partial v}{\partial t} + v\frac{\partial v}{\partial x}\right) = -\frac{\partial P}{\partial x}. \tag{8.12}$$

The evolution equation for the pressure is simply

$$\frac{dP(\rho)}{dt} = \frac{\partial P}{\partial t} + v\frac{\partial P}{\partial x} = c^2(\rho)\frac{d\rho}{dt} = c^2(\rho)\left[\frac{\partial \rho}{\partial t} + v\frac{\partial \rho}{\partial x}\right]. \tag{8.13}$$

Expanding these expressions in quasi-linear form (i.e. all partial derivatives involve only one quantity), we obtain

$$\begin{pmatrix} v & \rho & 0 \\ 0 & -v & -1/\rho \\ 0 & -\rho c^2 & -v \end{pmatrix} \begin{pmatrix} \rho_x \\ v_x \\ P_x \end{pmatrix} = \begin{pmatrix} \rho_t \\ v_t \\ P_t \end{pmatrix}. \tag{8.14}$$

The solution to the associated eigenvalue problem in λ is $\lambda = v, v+c, v-c$. These are the natural speeds at which information is carried. One is the medium's velocity and the others are the medium's velocity shifted in a positive and negative sense by the sound speed. The characteristic $\lambda = v$ corresponds to the material coordinate X – recall that $\rho \, dx = \rho_0 \, dX$. Hence the force equation along that characteristic $X(x, t)$ becomes

$$\frac{dv}{dt} = -\frac{1}{\rho}\frac{\partial P}{\partial x} = -\frac{1}{\rho_0}\frac{\partial P}{\partial X}. \tag{8.15}$$

Along the different characteristic lines, the evolution can be calculated in a similar way. We simply transform from coordinates x, t to s, t where s is the solution to $ds/dt = \lambda$. As a numerical method, characteristic schemes are rarely used, since the solutions evolve along mutually incompatible lines and extensive interpolation is required. However, they are particularly helpful in understanding the behavior around shocks and when troublesome boundary conditions are present. For example, the differential equation

$$\frac{\partial u}{\partial t} + \frac{\partial u}{\partial x} = 0 \tag{8.16}$$

has the general solution

$$u(x, t) = \mathcal{H}(x - t), \tag{8.17}$$

where $\mathcal{H}$ is any real-valued function and describes a right-going wave with a velocity of one. If we specify a boundary condition on the right-hand side of our domain, an inconsistency is almost certain to occur between the flow and the imposed boundary condition – to resolve that, we need to go back to the physics.

Having explored characteristic methods for quasi-linear first-order partial differential equations, let us return to the second-order linear partial differential equation. We will confine our focus to scalar situations – the vector problem can be resolved using diagonalization methods as described above. For convenience, we will focus on two independent variables – either two space variables or one space variable and time variable. Boundary data will be specified in one of several ways: either the dependent variable is prescribed on the boundary (Dirichlet) or its normal derivative is specified (Neumann). Homogeneous boundary data corresponds to a Dirichlet condition with vanishing dependent variable. Sometimes, mixed

Dirichlet–Neumann boundary conditions appear. Consider, then, the quasi-linear second-order equation

$$a \frac{\partial^2 u}{\partial x^2} + b \frac{\partial^2 u}{\partial x \partial y} + c \frac{\partial^2 u}{\partial y^2} = e. \tag{8.18}$$

For convenience, we will define

$$p \equiv \frac{\partial u}{\partial x}, \quad q \equiv \frac{\partial u}{\partial y}, \quad \alpha \equiv \frac{\partial^2 u}{\partial x^2}, \quad \beta \equiv \frac{\partial^2 u}{\partial x \partial y}, \quad \gamma \equiv \frac{\partial^2 u}{\partial y^2}. \tag{8.19}$$

The derivatives must satisfy

$$du = p\,dx + q\,dy$$
$$dp = \alpha\,dx + \beta\,dy$$
$$dq = \beta\,dx + \gamma\,dy, \tag{8.20}$$

and the differential equation provides a fourth relation

$$e = a\alpha + b\beta + c\gamma. \tag{8.21}$$

If we consider the latter equation plus the last two relations in the previous expressions as defining α, β, and γ, it follows that the solution exists and is unique unless the determinant of the matrix of coefficients vanishes, that is

$$\begin{vmatrix} dx & dy & 0 \\ 0 & dx & dy \\ a & b & c \end{vmatrix} = 0, \tag{8.22}$$

or

$$c\,(dx)^2 - b\,(dx)(dy) + a\,(dy)^2 = 0. \tag{8.23}$$

Note that this equation has the form of a *conic section*, and its determinant is coordinate invariant – according to the sign of the determinant (positive, zero, negative), the partial differential equation is regarded as elliptic, parabolic, or hyperbolic. When the latter condition holds, we have defined a characteristic line – there are two since this is a second-order system. When the latter condition is satisfied, no solution exists unless the other determinants of the system also vanish so that, to maintain sufficiency, we have

$$\begin{vmatrix} dp & dx & 0 \\ dq & 0 & dy \\ e & a & c \end{vmatrix} = 0, \tag{8.24}$$

or

$$e\,dx\,dy - a\,dp\,dy - c\,dq\,dx = 0. \tag{8.25}$$

These partial differential equation types are very relevant to physical processes. Wave equations are hyperbolic, diffusion equations are parabolic, and harmonic equations (e.g. Poisson's equation in electrostatics or gravitational potentials). All of these types of equations are important in the earth sciences.

The latter now defines the conditions that must be satisfied along the characteristics. Note, however, that initial data must be selected to lie along *both* characteristics; otherwise the specification of initial conditions is not complete. As an example, consider again the wave equation (8.1) which, from the above, has characteristics $x - t =$ constant and $x + t =$ constant (here, of course, we have replaced the role of y by t). Along these characteristics, we require respectively that $q - p =$ constant and $q + p =$ constant. Initial conditions must be given that influence both characteristics. For example, initial conditions that apply only when $x = t + x_0$ give us information along the $x + t$ characteristic, but only at one point on the $x - t$ characteristic.

One issue of profound importance regarding partial differential equations, say in x and t over a domain D, is whether there exist any conservation laws, i.e. quantities of the form

$$\mathcal{I}(t) \equiv \int_D \varphi \left[u(x, t) \right] dx, \tag{8.26}$$

which are *independent* of time. For example, the parabolic or diffusion equation

$$\frac{\partial u}{\partial t} = \frac{\partial^2 u}{\partial x^2}, \tag{8.27}$$

with $\partial u/\partial x = 0$ at the boundaries, conserves the integral $\int_D u(x, t) dx$. Conservation laws such as this, corresponding to the preservation of mass in the continuum problem together with any underlying symmetries, should be maintained if at all possible in numerical schemes. Having considered this brief overview of partial differential equations, we now turn our attention to their solution.

8.2 Survey of numerical methods

The partial differential equations of continuum systems are generally nonlinear and intractable by analytic methods, and numerical solutions are often essential. However, at the same time, it is essential that we guarantee that such approximate methods preserve the fundamental features of the problem. In addition to gross features such as conservation laws and symmetries, it is often important that we preserve the precise dynamical and geometrical characterization of a problem. Sometimes numerical methods are unstable. These present a major obstacle to the determination of the flow of a physically stable system. Similarly, one must guard against the possibility that a numerical method quenches an instability that is genuinely present in a physical problem. Thus, we must develop standards of consistency, accuracy, and stability as well as of conservation for any numerical method.

For simplicity, we will consider at least for hyperbolic and parabolic equations independent variables x and t. The essential problem is that we need to represent by a finite number of quantities the behavior of a continuum whose number of degrees of freedom is for all practical purposes infinite. Thus, it is convenient to think in terms of the statement of the finite difference problem where we employ samples of the dependent variable whose computational spacing remains finite. By consistency, we mean that in the limit $\Delta x \to 0$ and $\Delta t \to 0$ our numerical representation converges upon the true one. By accuracy, we mean the difference between the numerical approximation and the true expression in terms of a formal order parameter, i.e. a second order method has errors $O(\Delta x^2, \Delta t^2)$. Most computational schemes for partial differential equations are second order, but new methodologies are emerging that will permit fourth or even sixth order accuracy with only a modest, i.e. factor of two–four, increase in computational cost (Schiesser, 1991).

Finally, by stability we adopt the operational definition employed by the numerical analysis community: if the physical solution is not increasing (according to some mathematical norm), then the numerical solution also should not be increasing. It is important to note that a stable method need not be accurate. A physical behavior that damps need only preserve its non-growing state to be considered stable. On the other hand, an "accurate" method need not be stable. There is no contradiction here as accuracy is a short-term behavior while stability is a long-term characterization. Clearly, we desire stable and accurate methods. Moreover, as accuracy is also a local description, we wish to maintain any global features of the problem such as conservation laws.

Before beginning our discussion of finite difference methods, let us review for a moment the nature of spectral methods and of finite element methods. Recall, as an example, the parabolic differential equation (8.27) considered earlier and suppose now that it is valid over the domain $(0, 2\pi)$ where $u(0, t) = u(2\pi, t) = 0$. It is natural to consider $u(x, t)$ via a Fourier series

$$u(x, t) = \sum_{n=1}^{\infty} \eta_n(t) \sin\left(\frac{nx}{2}\right). \tag{8.28}$$

Since the Fourier series is a complete representation, it follows that we can describe any solution in this way. We observe, further, that we can convert the parabolic partial differential equation into an infinite set of ordinary differential equations in $\eta_n(t)$, namely

$$\frac{d\eta_n(t)}{dt} = -\frac{n^2}{4} \eta_n(t). \tag{8.29}$$

Each of these equations (or as many as we need to represent the initial data) can be solved by traditional schemes associated with ordinary differential equations. The description of a problem via a Fourier basis set is known as a "spectral representation" and is the basis of a *spectral method*. In other problems, other basis sets are more natural and schemes which employ them are referred to as Galerkin schemes after the Russian engineer who in 1915 first utilized them. Galerkin schemes are, therefore, a generalization of spectral methods. For example, some Galerkin schemes utilize orthogonal polynomials, e.g. *Chebyshev*, for their basis set in contrast with the Fourier series terms that form the basis set for spectral methods. This commonly occurs in simulations of the Earth's mantle. Hybrid methods, which are part finite difference and part Galerkin, are sometimes referred to as the *method of lines* (Schiesser, 1991). These methods seek to decompose partial differential equations into ordinary differential equations, as in the preceding equation, by exploiting the properties of an operator in an appropriate space to obtain such a separation. In nonlinear problems, this is not always easy to do but appears to offer, in many problems, the possibility of developing very accurate solutions.

Finite element methods, in contrast with boundary element methods discussed earlier, exploit either the variational principle implicit to continuum problems or some aspect of a Galerkin method, where integral properties rather than function values are employed over some "finite element" (Johnson, 2009). While in finite difference methods we evolve the dependent variable which is sampled at discrete points, in finite element methods we consider its integrated value on the interval between sampled points. Accordingly, equations then emerge which have the same flavor as the Galerkin decomposition above. We will not say anything further about this class of methods here.

Finite difference methods are among the most often used class of integration schemes. They are predicated upon the ideas that are at the foundation of differential calculus. However, rather than taking the limit of the relevant differential quantities, we in some sense maintain them in differential form. Consider, for example $u(x, t)$. It is convenient to consider u to be evaluated at discrete points $x_j = x_0 + j\Delta x$ and at discrete times $t_n = t_0 + n\Delta t$. Accordingly, we denote by u_j^n the quantity $u(x_j, t_n)$. As a simple example, consider once again the parabolic partial differential equation (8.27). We observe that

$$u_j^{(n+1)} = u_j^{(n)} + \frac{\Delta t}{\Delta x^2} \left[u_{j+1}^{(n)} - 2 u_j^{(n)} + u_{j-1}^{(n)} \right] + \mathrm{O}\left(\Delta t^2, \Delta x^2\right). \qquad (8.30)$$

The error term was obtained by simply taking a Taylor series expansion in both x and t around the point x_j at t_n. We refer to the spatial derivative term as being a "centered" difference as we are using information from each side of the current sample point. We refer to the time derivative term as being a "forward" difference as it takes us forward in time. Thus, we have established the consistency and accuracy of this method. Observe also that this method is *explicit*: the answer can be determined without requiring any knowledge of the desired solution. If we consider an oscillatory solution of the form

$$u_j^{(n)} = u^{(n)} \exp(i \, j \, \Delta x), \tag{8.31}$$

where $i = \sqrt{-1}$, this is equivalent to a Fourier analysis. Further, we note that all Fourier modes must damp since

$$\dot{u}_j^{(n)} = -j^2 u_j^{(n)}. \tag{8.32}$$

However, after a little bit of algebra, our finite difference expression (8.30) shows that

$$u_j^{(n+1)} = u_j^{(n)} \left[1 - \frac{4 \Delta t}{\Delta x^2} \sin^2\left(\frac{\Delta x}{2}\right) \right]. \tag{8.33}$$

Thus, we can assure stability if and only if

$$4 \Delta t < \Delta x^2. \tag{8.34}$$

This corresponds to the notion that the time step be selected to be less than the time it takes for a particle to diffuse over a distance of Δx. A more accurate but implicit integration scheme emerges if we handle the right-hand side of (8.33) in a different way. In particular, we can take the *average* of the $u_j^{(n)}$ term and the same term evaluated at the new time $n + 1$, i.e. $u_j^{(n+1)}$. We can show that the solution is stable for any choice of time step, if we are able to determine it. (We need to know the answer to find the answer; a simple iteration of the system rarely works. However, in many cases, the solution is accessible owing to the linearity of the problem. Interestingly, the implicit scheme is more accurate in this case as we have estimated the spatial second derivative at a time "intermediate" between n and $n + 1$.) Finally, suppose our boundary conditions remain such that the dependent variable vanishes on the boundary, say at points 0 and N. Then, we can readily show for either the explicit or implicit form that the quantity

$$I(n) = \sum_{j=1}^{N-1} u_j^{(n)} \tag{8.35}$$

never changes and therefore corresponds to the "mass conservation" law we considered earlier.

Hyperbolic equations which are related to wave propagation traditionally have been a source of problems in that they often are unstable. The essential problem here is that there is a directionality, associated with the characteristic lines, in which information is conveyed. This is the basis of the so-called "upwind" and "downwind" differencing schemes that are sometimes used. If c represents the fastest speed at which information travels, it generally becomes important (Richtmyer and Morton, 1967) to require that

$$c\,\Delta t < \Delta x. \tag{8.36}$$

This condition, usually known as the Courant–Friedrichs–Lewy or CFL condition, assures that information cannot travel in one time step further than one space step. One of the pioneering developments in the application of methods for treating hyperbolic systems is that due to Lax and to Wendroff, where they found a way to introduce explicitly the conservation laws into the formulation of the problem, as well as introduce some weak dissipation to quench any numerical instability that might occur. Further information on such methods lies beyond the scope of the present discussion and can be found in many textbooks (Richtmyer and Morton, 1967). More recently, the application of the method of lines (Schiesser, 1991) has allowed the development of higher-order integration schemes, especially methods that are fourth order in space and time.

Thus far, we have focused our discussion of finite difference methods on Eulerian schemes, where we remain bound to a fixed set of coordinates. Lagrangian schemes (Richtmyer and Morton, 1967) – at least for one-dimensional problems, as higher dimensions often introduce extreme conceptual difficulties – have also been of substantial use in computational explorations. Numerical instability also has been a hallmark of such problems, and artificial dissipation is often introduced to quench them. Further discussion lies beyond the scope of this chapter.

Finally, elliptic equations often lie at the heart of potential problems. For example, in two dimensions, the equation $\nabla^2 u(x, y) = 0$ can be approached by mixing a spectral representation in one dimension and a finite difference approximation in the other, e.g.

$$u(x, y) = \sum_{n=1}^{\infty} \mathcal{U}_n(y) \sin\left(\frac{nx}{2}\right), \tag{8.37}$$

and solve

$$\frac{\mathcal{U}_{n,j+1} - 2\mathcal{U}_{n,j} + \mathcal{U}_{n,j-1}}{\Delta y^2} = -\frac{n^2}{4}\mathcal{U}_{n,j}, \tag{8.38}$$

where j refers to the y variable. This can be solved directly since the matrix is "tridiagonal." Various forms of iterative schemes have emerged when we employ finite differences for all terms. Basically, we solve for the behavior in one direction, assuming we know the behavior in the opposite direction, and then reverse and repeat the process until convergence appears. This is the basis of various relaxation methods (*alternating direction implicit*) schemes. More complicated problems are also often amenable to iteration schemes, or may be cast in the form of a variational problem. The advantage of the latter is that global minimization methods can be adapted to the minimization of some quadratic form (e.g. *conjugate gradient* methods) and are highly efficient once implemented.

This concludes our brief overview of partial differential equations and their numerical solutions. This remains a developing field of investigation, and considerable room remains for substantial progress in computational continuum mechanics. Having provided this sketch of issues relating to computation, we turn now to the question of how nonlinearity becomes a fundamental issue in the dynamics of the Earth.

9

Nonlinearity in the Earth

As we scrutinize the landforms that surround us, we develop a sense of appreciation for the multitude of processes that shaped our planet. At the outset, we may think it is improbable that, out of this chaotic combination of physical processes, there could emerge any sense of abiding order. Nevertheless, we have come to appreciate during the preceding decades that the collective interaction of many different ingredients, as demonstrated by the Earth, yields manifestations of a new class of behavior which we refer to as "nonlinearity." It is important that we distinguish between the nonlinearity that we associate with the kinds of differential equations that we have discussed – i.e. the presence of terms that are of higher order than linear – from the collective behavior and self-organization that we sometimes observe. We reviewed this collective behavior in the context of solitary waves. Another aspect sometimes maintained by nonlinearity has roots in geometry and, as a consequence, influences in a profound way a variety of physical systems. Later, we shall survey features of percolation and fractal geometry as illustrations of this. We observed earlier how a "cascade" picture for energy transfer between different length scales at the same rate, as a form of self-organization, could help us understand turbulence and the emergence of power-law scalings. Perhaps this and other kinds of nonlinearity can help us understand the nature of earthquakes.

An important example of this kind of nonlinearity emerges from different statistics of earthquakes. Globally, earthquake statistics present power-law behavior that is consistent with scale-invariance, independent of location on the globe, the specific geometry of the earthquake fault involved, and the rheology of the associated geologic environment. The primary manifestation of this scale invariance is the Gutenberg–Richter law (Turcotte, 1997) which relates the earthquake magnitude M, essentially a logarithmic measure of the energy released, with the number of events N per unit time in a specified area with a magnitude greater than M. The empirical Gutenberg–Richter law has the form

$$\log_{10} N = -b M + \log_{10} \dot{a}, \tag{9.1}$$

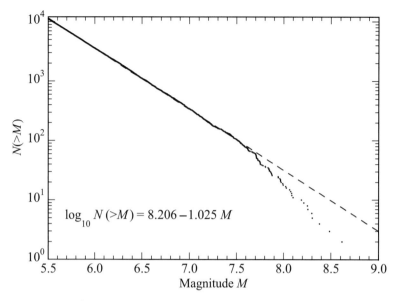

Figure 9.1 Global seismicity from 1977 to 2007.

where $\dot{a}$ and b are constants. While the rate $\dot{a}$ depends upon the geographic region and area considered and other factors, the value of b appears to be slightly above 1, essentially universally. To illustrate this remarkable result, we display in figure 9.1 all earthquake events above magnitude 5.5 observed globally during the years 1977–2007, inclusive. The data presented here comes from the Global Centroid Moment Tensor Project (CMT) catalog, which is available via the web.[1] While the observations spanned 31 years, we normalized the number count and fit the data using the prevailing standard of numbers of events above a given magnitude per year. We observe for this global data set that a "b-value" of approximately 1.025 emerges. Events above magnitude 7.5 fall in frequency below the Gutenberg–Richter relation (9.1), likely because the largest events are associated with ruptures that extend to significant depths below the surface where temperatures are much higher and the rocks are much more visco-elastic than at the surface. This figure suggests that the frequency–magnitude relationship becomes a steeper power law for higher magnitudes than at lower magnitudes. This figure, and the other statistical results cited below, are profound inasmuch as they establish universal scalings in seismic processes, a hallmark of nonlinearity.

A milestone development pertaining to our understanding of seismicity is the so-called "slider-block" model of Burridge and Knopoff (1967). Just as Reid (1911) developed his elastic rebound theory for earthquake events including the Great

---

[1] www.globalcmt.org/CMTsearch.html

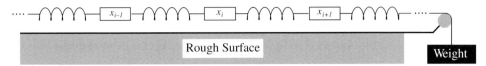

Figure 9.2 Burridge–Knopoff slider-block model.

San Francisco Earthquake, Burridge and Knopoff developed a skeletal mechanical model that mimicked these plate tectonic events and reproduced Gutenberg–Richter statistics. The Burridge–Knopoff model has, meanwhile, undergone major expansion by many investigators and has provided a fundamental enhancement in our theories of earthquake statistics.

Consider an ensemble of identical slider blocks connected by identical springs that rests upon a rough surface. Let us imagine that the center of mass of the ith block is situated at x_i and, when the blocks are at equilibrium with no forces acting, they are separated by a distance $\mathcal{D}$, and that Hooke's law governs the action of the springs and Amontons' Law governs friction of the sliders on the rough surface. The pulling of this chain of masses and springs by a steadily increasing weight, as shown in figure 9.2, sets into motion a complex set of events. According to Hooke's law, we can show that the net force acting on the ith block is proportional to $x_{i+1} - 2x_i + x_{i-1}$. Similarly, the potential energy stored in the system varies as $\sum (x_{i+1} - x_i)^2$. Initially all blocks are at rest until the block closest to the weight experiences sufficient force from the weight that it overcomes friction and is set into motion. Then the next block will also undergo slip, and so on. In so doing, however, the motion of the second and subsequent sliders alters the forces experienced by each of their neighbors and a complex system of events takes place. An "earthquake" happens when motion in an individual block initiates motion in connected blocks until the force experienced by each of the blocks falls below the threshold established by friction and all motion stops. Burridge and Knopoff (1967) monitored the potential energy in the system as a function of time, and observed a sawtooth-like curve with steady increases in potential energy being rapidly relieved by slip events although on average the potential energy would steadily increase. Burridge and Knopoff introduced many variations to their basic models, as have many others since, but the common feature to emerge from their investigation and those of others was a "frequency–magnitude" diagram not unlike figure 9.1 and a power-law or scale-invariant relationship between the frequency and energy involved (Turcotte, 1997).

Importantly, this kind of behavior can also be related to *percolation* (Stauffer and Aharony, 1994) and, possibly, *self-organized criticality* (Bak, 1996; Jensen, 1998). We shall discuss percolation processes and self-organized criticality later. Other empirical "laws" relating to seismicity are also evident in nature. Omori's law (1895) emerged following the 1891 Nobi earthquake (Scholz, 2002), and describes the frequency of aftershocks n following the main shock. Omori's law has the form

$$n = \frac{c}{(1 + t/t_0)^p}, \qquad (9.2)$$

where c is an observed rate constant, t_0 is a measured characteristic time and the power p is very close to 1. In addition to these two time-related statistical distributions, the spatial distribution of earthquakes also presents some power-law statistics (Kagan and Knopoff, 1980) that can be associated with *fractal* behavior (Mandelbrot, 1983; Peitgen et al., 1988; Feder, 1988). Kagan and Knopoff's result shows that earthquake faults form a complex network with fractal scaling. Large events occur on large faults while small events occur on small faults. There are many conjectures and models about how this takes place. For example, plate tectonics introduces large-scale stresses into a region resulting in microfracture (Allègre et al., 1982). Those microfractures in turn can fuse together to form larger ones, and so on in an inverse cascade culminating in large events (Newman and Knopoff, 1982, 1983; Knopoff and Newman, 1983). Other investigators have remarked on possible parallels between turbulence and earthquake events, and the relevance of power-law scalings and fractals (Gabrielov et al., 2000a,b). We will briefly describe fractal scaling later in this chapter. Other aspects of seismicity and nonlinearity are addressed by Scholz (2002), Kasahara (1981), Turcotte (1997), and Segall (2010) as well as in the seismology texts cited earlier (Ben-Menahem and Singh, 2000; Shearer, 2009; Aki and Richards, 2002; Bullen and Bolt, 1985).

The application of continuum mechanics to the mechanics of faulting – and, ultimately, to the study of earthquakes – remains one of the principal challenges of geophysics. This field is a blend of simple applications of the geometry of stress fields especially in two dimensions, the investigation of stress in the vicinity of "cracks" idealized as ellipses so that the tools of complex analysis can be exploited (Lawn, 1993), and a very rich phenomenology dealing with heterogeneous systems ("rocks" and large ensembles of Earth materials) with highly complicated alignments, force distributions, fluids, etc. Increasingly, there are important hints that nonlinearity is an essential ingredient in the earthquake mechanism. The nature of seismicity remains, perhaps, the most elusive goal of all. Here, we will try to represent some important themes in this critical field.

We will begin, following Lawn (1993) and Scholz (2002), by exploring the nature of friction and its phenomenology as ascertained by Amontons and by Coulomb as well as contemporary developments. We will then proceed to discuss the phenomenon of percolation, for example in lattice or crystalline structures, and how dramatic changes in material properties can result from only modest changes in the fraction of lattice squares that have failed. We will also briefly describe self-organized criticality theory (Bak, 1996; Jensen, 1998) and how these are relevant to models for earthquakes (Gabrielov et al., 1999). We will provide a short introduction to, and example of, fractals (Mandelbrot, 1983; Peitgen et al., 1988; Feder, 1988). We have already noted some simple environments wherein susceptibility to initial conditions can yield chaotic behavior (Turcotte, 1997; Drazin, 1992; Strogatz, 1994). Finally, we will conclude by presenting some summary comments on the role of nonlinearity emergent from continuum mechanics in the earth sciences.

9.1 Friction

The study of friction is one of the oldest, yet least well-understood, aspects of mechanical systems. Friction, in addition to fracture, is one of the central ingredients in the earthquake mechanism. Leonardo da Vinci in the fifteenth century was among the first to study friction in a systematic way. He discovered the two main laws of friction as well as observed that friction is less for smoother surfaces. His discoveries remained hidden in his *Codices* until they were rediscovered two centuries later by Amontons. Newton, in his *Principia*, made some fundamental quantitative observations about the nature of friction. Interestingly, frictional forces are in seeming violation, albeit not in actuality, of Newton's laws of motion. In 1699, Amontons stated the two main laws of friction, which we cited in section 2.7, namely that the frictional force is independent of the size of the surfaces in contact, and that friction is proportional to the normal load. Amontons also observed that the friction force is about one-third of the normal load, regardless of surface type or material. Rock friction, in contrast, is about twice as great.

In the century following Amontons, two fundamental developments emerged. First, the notion of *asperities* developed from the notion that protrusions on surfaces could interact, either rigidly or elastically, and that friction was due to the work done against gravity during the transport of material as the asperities were forced to ride up and over each other. Second, the difference between static and kinetic friction was also recognized. Coulomb used the asperity mechanism, albeit not calling it that, to describe how friction between static surfaces was greatest and would systematically diminish as the relative speed of the surfaces increased. The essential problem with these theories is that they fail to account for energy loss (Amontons' proposed mechanism was conservative) or for frictional wear.

In modern times, many investigators contributed to the development of the adhesion theory of friction, particularly for metals (Scholz, 2002). They envisioned all real surfaces as having topography, so that when two surfaces are brought together they only touch at a few points which they called asperities. The sum of all such contact areas is the real area of contact A_r, which is much smaller than the apparent or geometric area A of the contact surfaces. Moreover, it is only A_r, and not A, that is responsible for friction. They assumed, in response to an applied normal load N, that the contact area would grow, i.e. the interface would "yield" to the load, until

$$N = p A_r, \tag{9.3}$$

where p is the penetration hardness, a measure of the strength of the material. See, also, section 2.7; for simplicity, we are treating all quantities here as scalar variables. The frictional force F, meanwhile, would be the sum of the shear strength of the junctions

$$F = s A_r, \tag{9.4}$$

where s is the shear strength of the material, a feature which can be associated with molecular forces (adhesion). Combining these two equations, we can define the *coefficient of friction* μ by

$$\mu = \frac{F}{N} = \frac{s}{p}. \tag{9.5}$$

This latter expression is of fundamental importance as it describes friction according to the ratio of two different measures of strength (shear vs. penetration) for a given material. This description also explains the role of lubrication – s is reduced – although it does not do well at predicting the correct value for μ for a plastic material. For example, for ductile metals which have a yield strength in compression of σ_y, the penetration strength is $3\sigma_y$ while the shear strength is $\sigma_y/2$ which would predict $\mu = 1/6$ – which is a factor of two to four too low. This implies that other mechanisms, in addition to adhesion, are at work. In essence, a component of the "old" notion of friction must also apply. Real surfaces are "rough" and work must be done to overcome the interlocking of surface features – as proposed by Amontons. The recent discovery that surfaces are often "fractal" – i.e., have a texture which makes it necessary to describe it as having a dimension greater than two – must play an important role. Higher fractal dimension (complexity of surfaces) correlates with higher friction and could also help explain the factor of two difference between friction in rocks and metals. (We will discuss fractal geometry later in this chapter.) A further effect which must be introduced is the role of "gouge" – the physical break-up of irregularities on the surface, resulting in dissipation and producing a gritty material into the interface which can either reduce or enhance the resistance to shear. (This feature is also mirrored on a macroscopic scale with the formation of "damage zones" on earthquake faults.) The development of new microscopic investigation techniques plus new statistical and mathematical concepts, particularly fractals and scale invariance, offers the hope that substantial progress into the nature of friction will occur in coming years. We now turn our attention to the issue of fracture.

9.2 Fracture

The study of fracture remains one of the most intellectually challenging aspects of continuum mechanics, with very few analytic solutions known to the mathematical problems that emerge. On a microscopic level, fracture is intimately tied to the breaking of molecular bonds. There are two fundamental types of defects: cracks, which are surface defects, and dislocations, which are line defects. An essential feature of all defects is that they can produce concentrations of stress. These, in turn, are what provide the basis for earthquake events – just as defects produce stress concentrations, the localization of stress in excess of material strength can result in the production of defects.

The principal paradigm for fracture is provided by an elliptical hole (Lawn, 1993; Scholz, 2002), defined by

$$\frac{x^2}{c^2} + \frac{y^2}{b^2} = 1, \tag{9.6}$$

in a two-dimensional plate with a uniform tensile stress σ_∞ in the z-direction. From elasticity theory – see, for example, Narasimhan (1993) – the bottom and top of the hole have compressive stresses of magnitude $-\sigma_\infty$ and the left- and right-hand edges have tensile stress of magnitude

$$\sigma = \sigma_\infty \left[1 + 2\left(\frac{c}{b}\right)\right]. \tag{9.7}$$

Let us imagine that this crack is held open by the tensile stress and has a very small aspect ratio, i.e. $b \ll c$. This equation shows that the stress at the crack tip increases as a crack grows. (It is helpful to think of what is involved in trying to split apart a piece of wood. Once the crack exceeds some length, the stress at the crack tip exceeds the theoretical strength of the material, and the crack begins to grow at an accelerating pace.) What the former equation is telling us is that the tensile stress is being focused at each end due to the tiny radius of curvature and the "moment arm." The radius of curvature r of a curve $y(x)$ is given by

$$r^{-1} = \frac{|d^2y/dx^2|}{[1 + (dy/dx)^2]^{3/2}}. \tag{9.8}$$

After significant algebra, we find when $x = b$ that

$$r = \frac{c^2}{b}, \tag{9.9}$$

and, for $c \gg b$, that

$$\sigma \approx \sigma_\infty \cdot \sqrt{\frac{c}{r}}. \tag{9.10}$$

Further analysis shows that the stress field away from the crack tip varies according to $1/\sqrt{r}$, where r is now the distance from the crack tip – the inverse-square root scaling is an additional outcome of the analytic solution for this problem.

Griffith in the 1920s posed this problem at a more fundamental level, in the form of a thermodynamic energy balance for crack propagation and these ideas are reviewed by Scholz (2002). If the ellipsoidal crack of length $2c$ lengthens by an increment δc, then work W will be done by the external forces, and there will be a change in the internal strain energy U_e. Further, there will be an expenditure of energy in creating the new surfaces U_s (i.e. the extended crack). Thus, the total energy of the system, U, for a static crack will be

$$U = (-W + U_e) + U_s. \tag{9.11}$$

The combined term in parentheses is referred to as the "mechanical energy."

It follows that, if the cohesion between the incremental extension surfaces δc were removed, the crack would accelerate outward to a new lower energy configuration – thus, mechanical energy must decrease with crack extension. The surface energy, however, will increase with crack extension, because work must be done against the cohesion forces, i.e. breaking molecular bonds, in creating the new surface area.

The condition for equilibrium between these two competing influences is just

$$\frac{dU}{dc} = 0. \tag{9.12}$$

Griffith analyzed a rod under uniform tension, with length y, modulus E, and unit cross section loaded under a uniform tension σ. This has a strain energy $U_e = y\sigma^2/2E$, as we saw in section 5.2. If a crack of length $2c$ is introduced, the strain energy will increase by an amount $\pi c^2 \sigma^2/E$ so that

$$U_e = \frac{\sigma^2 \left(y + 2\pi c^2\right)}{2E}. \tag{9.13}$$

The rod becomes more compliant with the introduction of the crack, with an *effective modulus* $\mathcal{E}$ given by

$$\mathcal{E} = \frac{y}{y + 2\pi c^2} \cdot E. \tag{9.14}$$

The work done, therefore, in introducing the crack is

$$W = \sigma y \left(\frac{\sigma}{\mathcal{E}} - \frac{\sigma}{E}\right) = \frac{2\pi \sigma^2 c^2}{E}. \tag{9.15}$$

Meanwhile, the surface energy change is just

$$U_s = 4c\gamma, \tag{9.16}$$

where γ is the specific surface energy, the energy per unit area required to break the bonds. Hence, we have that

$$U(c) = -\frac{\pi c^2 \sigma^2}{E} + 4c\gamma. \tag{9.17}$$

Applying the equilibrium condition $dU(c)/dc = 0$, we obtain the critical stress at which a suitably oriented crack will be at equilibrium, namely

$$\sigma_{\text{Griffith}} = \sqrt{\frac{2E\gamma}{\pi c}}. \tag{9.18}$$

By relating this stress obtained from strictly thermodynamic considerations to that derived earlier, we note that the observed strength of materials must imply the existence of microscopic cracks, known as Griffith cracks. Further refinements to the theory have been made in recent years, inspired in many instances by the connection between solid state physics and continuum mechanics.

Finally, one last topic requires some discussion, the three crack propagation modes and the role of "stress intensity factors." Figure 9.3 illustrates the three modes of crack propagation. Mode I is the tensile or opening mode in which the crack wall displacements are normal to the crack. There are two shear modes. Mode II, in-plane shear, where the displacements are in the plane and normal to the crack edge; and, Mode III, antiplane shear, where the displacements are in the plane and parallel to the edge. The latter are analogous to edge and screw dislocations, respectively, in solid state physics.

If a crack is assumed to be planar and perfectly sharp, with no cohesion between the crack walls, near-field approximations to the crack-tip stress and displacement fields may be reduced to the simple analytic but approximate expressions:

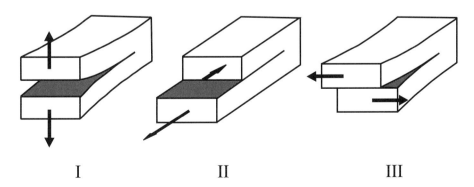

Figure 9.3 Geometry of crack mode propagation.

$$\sigma_{ij} = K_n (2 \pi r)^{-1/2} f_{ij}(\theta), \qquad (9.19)$$

and

$$u_i = \left(\frac{K_n}{2E}\right) \left(\frac{r}{2\pi}\right)^{1/2} f_i(\theta), \qquad (9.20)$$

where r is the distance from the crack tip and θ is the angle measured from the crack plane. The terms K_n are called the *stress intensity factors* for the mode n and the functions $f_{ij}(\theta)$ and $f_i(\theta)$ can be found in standard references (Lawn, 1993, pp. 53–54).

The theory for fracture, as well as for friction, has a large empirical component due to the complexity of the underlying continuum mechanical problem. Laboratory experiments remain a fundamental aspect of this subject and much remains to be done. Computer models for crack interaction and the evolution of large-scale fractures continue to hold promise in developing a better appreciation for the scalings that are observed. This field is basic to our understanding of faulting and the earthquake mechanism but there remain fundamental gaps in our understanding. New insights derived from laboratory experiment, such as those having to do with the fractal character of fault networks and fault surfaces as well as the self-similar character of damage zones and of rock gouge, will provide essential guidance for the work that has yet to be done.

9.3 Percolation and self-organized criticality

Earlier, we remarked that there exists a class of phenomena with geometrical roots that in many respects defy comprehension using straightforward extrapolation from day-to-day experience. Imagine, if you will, an array whose individual squares are either intact or have failed. So long as the failed squares do not link and connect to span the entire array, our system's integrity is assured. (We will consider squares that are mutually adjacent as being connected, and not those that are diagonally opposite each other.) Suppose further that we randomly and independently assign a probability of failure p between 0 and 1 to each square. Moreover, we assign a probability p_f, for $0 < p_f < 1$, for the array such that we call a square defective if its probability $p < p_f$ and intact if its probability $p > p_f$. If p_f was zero, then no squares will be defective. If p_f was one, then all squares will be defective. It is important to understand how this system behaves as p_f is gradually increased from 0 to 1.

We illustrate this in the figure 9.4 for four different values of p_f, namely 0.4, 0.5, 0.6, and 0.7. Each panel in this figure shows an array with 64×64 squares. Squares are displayed as being white, if intact, or black, if defective. When $p_f = 0.4$, we identify many small defective "pockets" or clusters of defective squares. As p_f is increased to 0.5, we observe that many such squares seem to merge.

9.3 Percolation and self-organized criticality

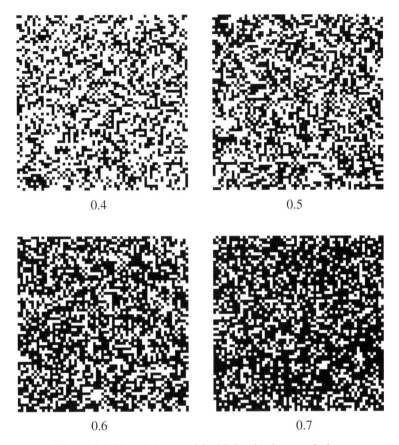

Figure 9.4 Transition to critical behavior in percolation.

However, when p_f becomes 0.6, something remarkable seems to have happened: in many instances, defective squares have linked to span the array from one edge to its diametrically opposite edge. As p_f is increased further, intact regions become smaller and increasingly isolated. It is now known that the linkage "percolates" when $p_f \approx 0.59275$ (Stauffer and Aharony, 1994; Turcotte, 1997). This is called the *critical probability* for percolation p_c.

In our portrayal of percolation, we have employed a description that reveals how this essentially geometrical phenomenon could be relevant to material failure and, by implication, earthquakes. Originally, percolation models were developed to help understand how contagious diseases in plants and animals could be controlled. Many years later, it was seen to help explain conductivity in materials constructed from an amalgam of conductors and insulators. However, the model we have just presented is fundamentally *static* and we would like to adapt it to dynamic circumstances.

Suppose we begin with an empty array, where all squares are intact. Then, at successive uniform intervals of time, we randomly select a square which then becomes defective. (If that particular square was already defective, then no change in its status occurs.) We repeat this process until one of the clusters of defective squares spans the entire system (Newman and Turcotte, 2002) as though percolation had occurred. In so doing, we have developed a *dynamic* percolation model. The original version of this model was developed to understand better the statistics of forest-fires. We replace the random selection of defective squares with the planting of trees in what will become a "forest." (We presume that nothing happens if we attempt to plant a tree on an already-occupied square.) Our metaphor becomes complete when we allow for "fires" to occur by the rare insertion of a lit match on a randomly selected square. If that square is vacant, then nothing happens. If that square is occupied, i.e. has a tree present, then that tree and all others that are mutually adjacent to it, i.e. all trees in that cluster, will burn. We further assume when a tree burns that it leaves behind a vacant square that can accept a new planting in the future. In this way, this dynamic *forest-fire* model (Turcotte, 1997) develops a steady rate of forest-fire production, once the system equilibrates, and its frequency–magnitude diagram (where the magnitude is the logarithm of the area consumed in a fire) looks remarkably like a Gutenberg–Richter plot with

$$\log_{10} N (> A) = -\log_{10} A + \text{constant},$$

i.e., having a "b-value" near -1. The percolation models we have just discussed, both dynamic and static, allow for only two states at each square. We will now discuss a class of models where this restriction is eliminated.

Suppose now that we deposit a unit of stress on a randomly selected square in our array with the passage of a unit of time. Further, we will assume that each square, upon accumulating four units of stress will release to its four adjacent squares 1 unit each of stress. Moreover, if any of those squares thereby achieves four units of stress, those four units will also be redistributed to the adjoining four squares, and so on until the changes to the array for a given deposit of stress ceases. Whenever the initial deposit of a unit of stress results in a redistribution, we say that we have a model "earthquake." The total number of stress units distributed, both from the original and any further redistribution of stress triggered as a consequence, constitutes the size of the event. After some length of time, the system settles down and maintains a kind of dynamical equilibrium with a well-defined average rate of events. The system size does interfere with the problem, as it breaks the scale-invariance, but otherwise we obtain power-law statistics in this situation although it breaks down at very large event size. This process has been given the name *self-organized criticality* (SOC) by Bak (1996) and his coworkers; Jensen (1998) provides a comprehensive review of the subject. While some have argued that the forest-fire model is self-organized critical, Gabrielov *et al.* (1999) showed otherwise and developed an explicit derivation of its frequency–magnitude law. The important lesson to be drawn from this elaboration of percolation ideas is that geometry alone can be responsible for behaviors that exhibit power-law behavior in the absence of other physically imposed length scales. Moreover, these processes where one scale length interacts with successively smaller (or larger) scale lengths in a scale-independent way could help describe many different geophysical phenomena. We turn now to the nature of fractals, as the *sine qua non* of nonlinearity and scale-invariant processes.

9.4 Fractals

We are accustomed to thinking of quantities such as the length of an object as lacking in controversy. In the real world, however, that question is deceptively simple. Lewis Frye Richardson speculated that the answer could depend on the length of ruler employed. He assembled data for the length of the coastlines of a number of countries and then emerged with the remarkable conclusions that the answer depended on the length of the ruler used and that our intuitive notion of one, two, or three dimensions was inadequate (Mandelbrot, 1983; Peitgen *et al.*, 1988).

In some sense, we are confronting the problem of calculating the length of an object with some measure of texture or roughness. We have already referred to how, in nature, we see situations where in a given problem different length scales interact with each other in a special way. To make this point, we will begin with a special example, known as the *triadic Koch curve* or "snowflake." We construct this object, as shown in figure 9.5, by drawing first a straight line. Then, we extract its middle third and create the edges of a "tent" using line segments of the same length. So, in our figure, we go from level 0 to level 1 with one line segment of length ℓ being replaced by four segments of length $\ell/3$ and with the accompanying change in the shape of what was once a straight line. In each succeeding level, we take an existing line segment, and replace it by four segments that are $1/3$ as long. This can be described in simple terms by assigning a length ℓ_i to each of the levels denoted by i. Thus, we observe

$$\ell_0 = \ell \tag{9.21}$$

and

$$\ell_{i+1} = \ell_i/3 \quad \text{or} \quad \ell_i = 3^{-i}\ell_0 \quad \text{for} \quad i = 0, 1, \ldots . \tag{9.22}$$

The number of "rulers" needed at each of these levels, say N_i is described by

$$N_i = 4^i, \quad \text{for} \quad i = 0, 1, \ldots . \tag{9.23}$$

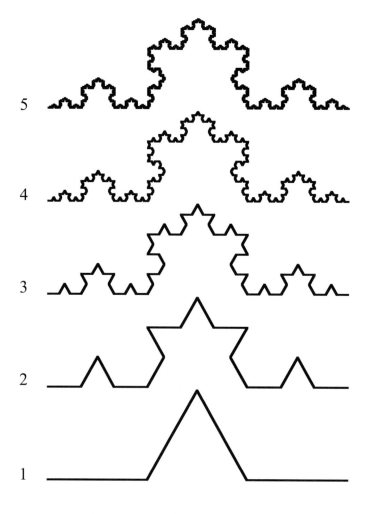

Figure 9.5 Fractal geometry in a Koch snowflake.

Accordingly, the length L_i at each level i that we can measure is given by

$$L_i = N_i \ell_i = \left(\frac{4}{3}\right)^i \ell. \qquad (9.24)$$

Normally, we expect that $N \propto \ell^{-1}$. Here, instead, we find after some elementary algebra that

9.4 Fractals

$$N_i = N_0 \left(\frac{\ell_i}{\ell_0}\right)^{-D}, \tag{9.25}$$

where we will call D the *fractal dimension* (Mandelbrot, 1983; Peitgen *et al.*, 1988; Jensen, 1998), which turns out to be

$$D = \frac{\ln 4}{\ln 3} \approx 1.26 > 1. \tag{9.26}$$

This process of iteration is intimately related to the methodology of *renormalization*.

It is natural to ask how we can obtain a non-integer dimensionality. This feature emerges because the object that we are trying to describe has a texture, and that it injects part of its character into the second dimension. In a qualitative sense, we can identify higher fractal dimensions with increased roughness or fuzziness. Mandelbrot (1983) coined the term "fractal," coming from the Latin word *fractus* meaning "to break creating irregular fragments." Mandelbrot's concept of the fractal dimension has stood the test of time inasmuch as it provides a useful methodology for characterizing objects whose dimensionality was not clear. This process of "ruler-counting" provides a convenient device to assess the dimensionality of an object that interpolates in some way between one and two dimensions.

Several features related to fractals can be associated with this demonstration.

(1) We employed a process of *iteration* to generate the successive levels of this figure, in a *renormalization* process.
(2) Our method of generation could be related to a *direct cascade* proceeding from the largest to the smallest scales.
(3) At every level, what emerged appeared the same in a statistical sense as what emerged at any other level – we say that the process is *self-similar*.
(4) In the limit of an infinite number of levels, we are confronting a process that is fundamentally *non-differentiable*.
(5) Our methodology for generating this fractal was *deterministic*, i.e. followed specified rules without alteration.

At the same time, we note that we can adapt these rules by introducing certain random components into the iteration process, thereby creating a *stochastic fractal*. For example, in the above, we could have removed a segment from somewhere in the "middle" of the line, where the length and center of that segment would be established according to some probability rule. Similarly, in constructing the "tent," we could have employed line segments with some degree of randomness in the selection of their lengths. Also, we could have elected to have a fraction of the tents pointed down. For example, if we had divided our initial line segment into four segments with randomly selected lengths, and transformed our original line into an object with eight randomly chosen line segments, we could have developed a fractal with dimensionality of $\ln 8/\ln 4 = 3/2$, which is what would be observed in the Brownian motion associated with the kinetic theory of gases. It is notable that many objects in nature satisfy our definition of stochastic fractals. Moreover, using these rules for producing fractals, we can now generate images of artificial worlds that appear remarkably realistic, inasmuch as our visual processing has adapted to exploit self-similarity. Fourier methods are widely used to generate fractal images (Turcotte, 1997; Peitgen *et al.*, 1988), unlike the scheme we used to generate the Koch snowflake. Mandelbrot (1983) incorporates a number of images of artificial planets, landscapes, and islands that have become fixtures of Hollywood. Our appreciation for this brand of complexity has now caught up with what our eyes now take for granted.

The survey we have provided in this chapter barely scratches the surface of nonlinearity and its manifestations on the Earth. Earlier, we noted the potential for *chaos* emerging from differential equations such as the Lorenz model (6.1) for convection that yield behaviors, according to the initial conditions chosen, that can depart exponentially rapidly from the original choice of initial conditions. We characterized situations where there were large numbers of degrees of freedom, as would be necessary in describing a fluid, as producing *complexity*. Nevertheless, in the face of these chaos-driven behaviors, we also witnessed the potential in chapter 6 for collective modes of behavior as manifest in *solitons* and other *solitary waves*. We then considered issues pertinent to the scaling properties in these complex systems. In situations containing no built-in length scale, the behaviors that obtained could be largely *scale invariant* and possess *power-law* scalings, as we also saw in the context of fluid mechanics and turbulence. We then observed that geometry could play a dominant role with special, sometimes counter-intuitive behaviors being present, such as *percolation* which conferred on a random system the ability to manifest *self-organization* if certain conditions were met. Moreover, *percolation* can also be related to phase transitions in materials and may play an important role in the Earth. *Self-organized criticality* can help explain the specific power-laws observed in the Earth without requiring substantial attention to detail in their execution. Remarkably, skeletal mechanical models containing the fundamental elements of faults and fault networks can now reproduce many of the different kinds of earthquake statistics and scalings that have been observed. All told, the Earth remains an amazing place!

References

Ablowitz, M. J., and Segur, H. (1981). *Solitons and the Inverse Scattering Transform*. Vol. 4. Philadelphia: SIAM.

Aki, K., and Richards, P. G. (2002). *Quantitative Seismology*. 2nd edn. Sausalito, CA: University Science Books.

Allègre, C. J., Le Mouel, J. L., and Provost, A. (1982). Scaling rules in rock fracture and possible implications for earthquake prediction. *Nature*, **297** (May), 47–49.

Arfken, G. B., and Weber, H. J. (2005). *Mathematical Methods for Physicists*. 6th edn. Burlington, MA: Elsevier Academic Press.

Aris, R. (1989). *Vectors, Tensors, and the Basic Equations of Fluid Mechanics*. New York: Dover Publications.

Bak, P. (1996). *How Nature Works: the Science of Self-organized Criticality*. New York: Copernicus.

Batchelor, G. K. (1953). *The Theory of Homogeneous Turbulence*. Cambridge: Cambridge University Press.

Batchelor, G. K. (1967). *An Introduction to Fluid Dynamics*. Cambridge: Cambridge University Press.

Ben-Menahem, A., and Singh, S. J. (2000). *Seismic Waves and Sources*. 2nd edn. Mineola, NY: Dover Publications.

Boas, M. L. (2006). *Mathematical Methods in the Physical Sciences*. 3rd edn. Hoboken, NJ: Wiley.

Bullen, K. E, and Bolt, B. A. (1985). *An Introduction to the Theory of Seismology*. 4th edn. Cambridge: Cambridge University Press.

Burridge, R., and Knopoff, L. (1967). Model and theoretical seismicity. *Bulletin of the Seismological Society of America*, **57**(3), 341–371.

Butt, R. (2007). *Introduction to Numerical Analysis using MATLAB*. Hingham, MA: Infinity Science Press.

Carmichael, R. S. (1989). *Practical Handbook of Physical Properties of Rocks and Minerals*. Boca Raton, FL: CRC Press.

Carmo, M. P. do (1976). *Differential Geometry of Curves and Surfaces*. Englewood Cliffs, NJ: Prentice-Hall.

Chadwick, P. (1999). *Continuum Mechanics: Concise Theory and Problems*. 2nd corrected and enlarged edn. Mineola, NY: Dover Publications, Inc.

Chandrasekhar, S. (1961). *Hydrodynamic and Hydromagnetic Stability*. New York: Dover Publications.

Chandrasekhar, S. (1995). *Newton's Principia for the Common Reader*. Oxford: Clarendon Press.
Cole, J. D. (1951). On a quasilinear parabolic equation occurring in aerodynamics. *Quarterly of Applied Mathematics*, **9**, 225–236.
Drazin, P. G. (1992). *Nonlinear Systems*. Cambridge: Cambridge University Press.
Drazin, P. G. (2002). *Introduction to Hydrodynamic Stability*. Cambridge: Cambridge University Press.
Drazin, P. G., and Johnson, R. S. (1989). *Solitons: an Introduction*. Cambridge: Cambridge University Press.
Drazin, P. G., and Reid, W. H. (2004). *Hydrodynamic Stability*. 2nd edn. Cambridge: Cambridge University Press.
Dummit, D. S., and Foote, R. M. (2004). *Abstract Algebra*. 3rd edn. Hoboken, NJ: Wiley.
Faber, T. E. (1995). *Fluid Dynamics for Physicists*. Cambridge: Cambridge University Press.
Feder, J. (1988). *Fractals*. New York: Plenum Press.
Fowler, A. (2011). *Mathematical Geoscience*. 1st edn. Interdisciplinary applied mathematics, vol. 36. New York: Springer.
Fox, L. (1962). *Numerical Solution of Ordinary and Partial Differential Equations: Based on a Summer School held in Oxford, August–September 1961*. Proceedings of summer schools organised by the Oxford University Computing Laboratory and the Delegacy for Extra-mural Studies, vol. 1. Oxford: Pergamon Press.
Fröberg, C. E. (1985). *Numerical Mathematics: Theory and Computer Applications*. Menlo Park, CA: Benjamin/Cummings Pub. Co.
Fung, Y. S. (1965). *Foundations of Solid Mechanics*. Englewood Cliffs, NJ: Prentice-Hall, Inc.
Gabrielov, A., Newman, W. I., and Turcotte, D. L. (1999). Exactly soluble hierarchical clustering model: inverse cascades, self-similarity, and scaling. *Physical Review E*, **60** (Nov.), 5293–5300.
Gabrielov, A., Zaliapin, I., Newman, W. I., and Keilis-Borok, V. I. (2000a). Colliding cascades model for earthquake prediction. *Geophysical Journal International*, **143** (Nov.), 427–437.
Gabrielov, A., Keilis-Borok, V., Zaliapin, I., and Newman, W. I. (2000b). Critical transitions in colliding cascades. *Physical Review E*, **62** (July), 237–249.
Gantmakher, F. R. (1959). *The Theory of Matrices*. New York: Chelsea Pub. Co.
Gear, C. W. (1971). *Numerical Initial Value Problems in Ordinary Differential Equations*. Prentice-Hall series in automatic computation. Englewood Cliffs, NJ: Prentice-Hall.
Ghil, M., and Childress, S. (1987). *Topics in Geophysical Fluid Dynamics: Atmospheric Dynamics, Dynamo Theory, and Climate Dynamics*. Applied mathematical sciences, vol. 60. New York: Springer-Verlag.
Gill, A. E. (1982). *Atmosphere–Ocean Dynamics*. New York: Academic Press.
Goldstein, H., Poole, C. P, and Safko, J. L. (2002). *Classical Mechanics*. 3rd edn. San Francisco: Addison Wesley.
Gurtin, M. E. (1981). *An Introduction to Continuum Mechanics*. Vol. 158. New York: Academic Press.
Helmholtz, H. von (1868). On the facts underlying geometry. *Abhandlunger der Königlicher Gesellschaft der Wissenschafter zu Göttinger*. Vol. 15.
Higham, N. J. (2002). *Accuracy and Stability of Numerical Algorithms*. 2nd edn. Philadelphia: Society for Industrial and Applied Mathematics.
Holton, J. R. (2004). *An Introduction to Dynamic Meteorology*. 4th edn. Vol. 88. Burlington, MA: Elsevier Academic Press.

Hopf, E. (1950). The partial differential equation $u_t + uu_x = \mu_{xx}$. *Communicators or Pure and Applied Mathematics*, **3**, 201–230.

Houghton, J.T. (2002). *The Physics of Atmospheres*. 3rd edn. Cambridge: Cambridge University Press.

Jackson, J. D. (1999). *Classical Electrodynamics*. 3rd edn. New York: Wiley.

Jensen, H. J. (1998). *Self-organized Criticality: Emergent Complex Behavior in Physical and Biological Systems*. Vol. 10. Cambridge: Cambridge University Press.

Johnson, C. (2009). *Numerical Solution of Partial Differential Equations by the Finite Element Method*. Dover books on mathematics. Mineola, NY: Dover Publications.

Kagan, Y. Y., and Knopoff, L. (1980). Spatial distribution of earthquakes: the two-point distribution problem. *Geophysical Journal*, **62**, 303–320.

Kasahara, K. (1981). *Earthquake Mechanics*. Cambridge Earth Science Series. Cambridge: Cambridge University Press.

Kennett, B. L. N. (1983). *Seismic Wave Propagation in Stratified Media*. Cambridge: Cambridge University Press.

Kincaid, D., and Cheney, E. W. (2009). *Numerical Analysis: Mathematics of Scientific Computing*. 3rd edn. The Sally series, vol. 2. Providence, RI: American Mathematical Society.

Knopoff, L., and Newman, W. I. (1983). Crack fusion as a model for repetitive seismicity. *Pure and Applied Geophysics*, **121** (May), 495–510.

Landau, L. D., and Lifshitz, E. M. (1987). *Fluid Mechanics*. 2nd edn. Course of Theoretical Physics, vol. 6. Oxford, England: Pergamon Press.

Landau, L. D., Lifshitz, E. M., Kosevich, A. M., and Pitaevskiĭ, L. P. (1986). *Theory of Elasticity*. 3rd edn. Course of Theoretical Physics, vol. 7. Oxford: Pergamon Press.

Lawn, B. R. (1993). *Fracture of Brittle Solids*. 2nd edn. Cambridge: Cambridge University Press.

Mandelbrot, B. B. (1983). *The Fractal Geometry of Nature*. Updated and augmented edn. New York: W. H. Freeman.

Marshall, J., and Plumb, R. A. (2008). *Atmosphere, Ocean, and Climate Dynamics: an Introductory Text*. Vol. 93. Amsterdam: Elsevier Academic Press.

Mase, G. E., and Mase, G. T. (1990). *Continuum Mechanics for Engineers*. Boca Raton, FL: CRC Press.

Mathews, J., and Walker, R. L. (1970). *Mathematical Methods of Physics*. 2nd edn. New York: W. A. Benjamin.

McKelvey, J. P. (1984). Simple transcendental expressions for the roots of cubic equations. *American Journal of Physics*, **52**(3), 269–270.

Millman, R. S., and Parker, G. D. (1977). *Elements of Differential Geometry*. Englewood Cliffs, NJ: Prentice-Hall.

Morse, P. M., and Feshbach, H. (1953). *Methods of Theoretical Physics*. International series in pure and applied physics. New York: McGraw-Hill.

Müser, M. H., Wenning, L., and Robbins, M. O. (2001). Simple microscopic theory of Amontons's laws for static friction. *Physical Review Letters*, **86**(7), 1295–1298.

Narasimhan, M. N. L. (1993). *Principles of Continuum Mechanics*. New York: Wiley.

Newman, W. I. (2000). Inverse cascade via Burgers equation. *Chaos*, **10** (June), 393–397.

Newman, W. I., and Knopoff, L. (1982). Crack fusion dynamics: A model for large earthquakes. *Geophysical Research Letters*, **9**, 735–738.

Newman, W. I., and Knopoff, L. (1983). A model for repetitive cycles of large earthquakes. *Geophysical Research Letters*, **10** (Apr.), 305–308.

Newman, W. I., and Turcotte, D. L. (2002). A simple model for the earthquake cycle combining self-organized complexity with critical point behavior. *Nonlinear Processes in Geophysics*, **9**, 453–461.

Nickalls, R. W. D. (1993). A new approach to solving the cubic; Cardan's solution revealed. *The Mathematical Gazette*, **77**(480), 354–359.

Oertel, G. F. (1996). *Stress and Deformation: a Handbook on Tensors in Geology*. New York: Oxford University Press.

Pedlosky, J. (1979). *Geophysical Fluid Dynamics*. New York: Springer Verlag.

Peitgen, H.-O., Saupe, D., and Barnsley, M. F. (1988). *The Science of Fractal Images*. New York: Springer-Verlag.

Peyret, R. (2000). *Handbook of Computational Fluid Mechanics*. San Diego, CA: Academic Press.

Peyret, R., and Taylor, T. D. (1990). *Computational Methods for Fluid Flow*. Corr. 3rd print edn. New York: Springer-Verlag.

Pope, S. B. (2000). *Turbulent Flows*. Cambridge: Cambridge University Press.

Press, F., and Siever, R. (1986). *Earth*. 4th edn. New York: W. H. Freeman.

Reid, H. F. (1911). The elastic-rebound theory of earthquakes. *Bulletin of the Department of Geology, University of California Publications*, **6**(19), 413–444.

Richtmyer, R. D., and Morton, K. W. (1967). *Difference Methods for Initial-value Problems*. 2nd edn. Interscience tracts in pure and applied mathematics, vol. 4. New York: Interscience Publishers.

Schiesser, W. E. (1991). *The Numerical Method of Lines: Integration of Partial Differential Equations*. San Diego: Academic Press.

Scholz, C. H. (2002). *The Mechanics of Earthquakes and Faulting*. 2nd edn. Cambridge: Cambridge University Press.

Schubert, G., Turcotte, D. L., and Olson, P. (2001). *Mantle Convection in the Earth and Planets*. Cambridge: Cambridge University Press.

Schutz, B. F. (1980). *Geometrical Methods of Mathematical Physics*. Cambridge: Cambridge University Press.

Segall, P. (2010). *Earthquake and Volcano Deformation*. Princeton, NJ: Princeton University Press.

Segel, L. A., and Handelman, G. H. (1987). *Mathematics Applied to Continuum Mechanics*. New York: Dover Publications.

Shearer, P. M. (2009). *Introduction to Seismology*. 2nd edn. Cambridge: Cambridge University Press.

Sleep, N. H., and Fujita, K. (1997). *Principles of Geophysics*. Malden, MA: Blackwell Science.

Spencer, A. J. M. (1980). *Continuum Mechanics*. Longman mathematical texts. London: Longman.

Stauffer, D., and Aharony, A. (1994). *Introduction to Percolation Theory*. Rev., 2nd edn. London: Taylor and Francis.

Strogatz, S. H. (1994). *Nonlinear Dynamics and Chaos: With Applications to Physics, Biology, Chemistry, and Engineering*. Reading, MA: Addison-Wesley Pub.

Tennekes, H., and Lumley, J. L. (1972). *A First Course in Turbulence*. Cambridge, MA: MIT Press.

Turcotte, D. L. (1997). *Fractals and Chaos in Geology and Geophysics*. 2nd edn. Cambridge: Cambridge University Press.

Turcotte, D. L., and Schubert, G. (2002). *Geodynamics*. 2nd edn. Cambridge: Cambridge University Press.

Vallis, G. K. (2006). *Atmospheric and Oceanic Fluid Dynamics: Fundamentals and Large-scale Circulation*. Cambridge: Cambridge University Press.

Van Dyke, M. (1982). *An Album of Fluid Motion*. Stanford, CA: The Parabolic Press.

Whitham, G. B. (1974). *Linear and Nonlinear Waves*. Pure and applied mathematics. New York: Wiley.

Index

acceleration field, 73
accuracy condition, 154
adiabatic, 94
adiabatic fluid, 117
alternating direction implicit, 157
Amontons, 41, 42, 161–163
angular momentum, 20, 75
asperities, 41, 163

balance laws, 68
barotropic fluid, 117
Bernoulli's principle, 120
biharmonic equation, 104
body force, 28
boundary conditions, 81
Boussinesq approximation, 139
Brunt–Vaisala frequency, 141, 143
bulk modulus, 91
Burgers vector, 108

Cauchy deformation tensor, 54
Cauchy stress princple, 29
Cauchy stress tensor, 31
Cayley–Hamilton theorem, 15
chaos, 112, 174
Chebyshev polynomial, 155
closure problem, 82, 124
coefficient of bulk viscosity, 115
coefficient of compression, 93
coefficient of extension, 95
coefficient of friction, 42
coefficient of friction μ, 164
complexity, 113
compressive stress, 32
confocal quadrics, 67
conjugate gradient methods, 158
conservation laws, 153
conservation of mass law, 71
consistency condition, 154

constitutive equations, 69, 79
continuity equation, 72
convective derivative, 53
Coriolis force, 20, 139
Coriolis parameter, 142
Coulomb, 41, 162, 163
Courant–Friedrichs–Lewy CFL condition, 157
Cramer's Rule, 63
creep, 81
critical probability, 169
crystalline structure, 106
cubic polynomial, 37
curvature, 18
cutting plane, 29
cyclonic flow, 142

deformation, 49
deformation gradient tensor, 54
deformed configuration, 51
density, 28, 69
deviator stress, 45
diffusion equation, 127, 148
dimensional analysis, 32, 135
Dirac delta function, 103
direct cascade, 124
Dirichlet boundary condition, 151
dispersion, 128
displacement, 49, 50
dissipation subrange, 125
downwind difference, 157
dyad, 5

earthquake, 41
eddy viscosity, 124
eigenvalue, 12
eigenvector, 12
Einstein summation convention, 1
Ekman number, 140
elastic modulus tensor, 107

elastic oscillations, 101
elastic rebound theory, 41
elastic waves, 101
elliptic equation, 148
energy equation, 78
engineering shear strain, 56
entropy, 68
equation of state, 81
equilibrium conditions, 33
equilibrium equations, 74
Eulerian, 52
extremal stress, 39

Fick's law, 78
field equations, 68
finite difference methods, 68
finite element method, 68, 155
first octant, 44
first rank tensor, 1
fluctuation, 121
forest-fire model, 170
Fourier's law, 78
fractal, 162, 164
fractal dimension, 173
friction, 41, 163
fundamental theorem of algebra, 13

Galerkin method, 155
geopotential, 139
geostrophic, 140
geostrophic approximation, 142
GFD, 134
Gibbs free energy, 85
gouge, 164
Gram–Schmidt orthogonalization, 23
Green's deformation tensor, 54
Green's function, 103, 128

heat flux vector, 78
Helmholtz, 49
Helmholtz free energy, 85
Helmholtz's Decomposition Theorem, 102
Hermitian, 8
homogeneity, 27
homogeneous deformation, 94
Hooke's law, 93
hydrostatic, 118
hydrostatic compression, 85, 91
hydrostatic equilibrium equation, 141
hydrostatic pressure, 114
hydrostatic state, 45
hyperbolic equation, 148, 156

identity matrix, 9
ill-conditioned, 147
incompressible fluid, 117
inelastic, 89
inertial subrange, 125

inertial terms, 74
infinitesimal deviator strain tensor, 57
infinitesimal rotation tensor, 58
infinitesmial spherical strain tensor, 57
internal energy, 77
inverse cascade, 125, 128
inviscid fluid, 118
irrotational, 119
isothermal, 94
isothermal fluid, 117
isothermal modulus, 98
isotropic, 27

Jacobian, 52
jump conditions, 127

KdV, 128
Kelvin's circulation theorem, 122
kinematic viscosity, 136
kinetic energy, 76
Koch curve, 171
Kolmogorov–Obukhov law, 125
Korteweg–de Vries, 128
Kronecker delta function, 2

Lagrange multiplier method, 39
Lagrangian, 50, 52
Lagrangian finite strain tensor, 54
Lamé coefficients, 90
laminar flow, 117
left stretch tensor, 59
Levi-Civita permutation symbol, 2
linear elastic solid, 79
linear momentum, 73
linear viscous fluid, 80
local equations of motion, 73
local equilibrium conditions, 33
longitudinal strain, 55
Lorenz model, 112

mapping, 49
matrix, 7
matrix function, 14
maximum stress, 39
Maxwell relations, 86, 89
mean normal strain, 57
method of characteristics, 148, 149
method of lines, 155
metric tensor, 67
minimum stress, 39
modulus of compression, 91
modulus of extension, 95
modulus of rigidity, 91
Mohr's circles, 43
moment equations, 124
moment of momentum, 75

Navier–Stokes, 117

Neumann boundary condition, 151
Newton identities, 14
Newtonian fluid, 80
normal strain, 55

Obukhov's law, 125
octahedral shear stress, 46
olivine, 57

P wave, 103
parabolic equation, 148
Pascal, 32
percolation, 161, 162
perfect fluid, 118
Piola–Kirchoff stress tensor, 74
plane strain, 57
plane stress, 44
plastic material, 80
Poisson's ratio, 95
polar decomposition, 59
potential equation, 148
potential flow, 120
Prandtl number, 113, 142
pressure scale height, 141
primary wave, 103
principal direction, 34
principal stress, 34
principle of linear momentum, 73
production subrange, 125
pure shear, 91

quadratic form, 8
quasi-geostrophic, 142

Rankine–Hugoniot conditions, 127
rate of deformation tensor, 59
rate of stretching per unit stretch, 61
Rayleigh number, 113, 142
Rayleigh–Bénard convection, 142
Rayleigh–Taylor instability, 141
renormalization, 173
Reynolds number, 136
right stretch tensor, 59
rigid body displacement, 49
Rossby number, 140
rotation, 9, 16
rotation vector, 58

S wave, 103
scale-invariance, 159
screw dislocation, 107
second law of thermodynamics, 68, 78
second rank tensor, 1
secondary wave, 103
self-organized criticality, 161, 170
self-similarity, 125
Serret–Frenet, 20
shear modulus, 91
shear rate, 61

shock, 126
shock front, 126
simple extension, 94
SOC, 170
solitary waves, 128
solitons, 128
spatial velocity gradient, 59
spectral method, 68, 155
spherical state of stress, 45
spin tensor, 60
spinel, 57
stability condition, 154
steady flow, 118
Stokesian flow, 115
strain tensor, 50, 51
streamlines, 122
stress function, 106
stress intensity factor, 168
stress power, 77
stress work, 77
stretch ratio, 58
surface force, 28

Taylor–Proudman theorem, 122, 123
tensile stress, 32
tensor, 1
thermal conductivity, 78
thermal energy balance, 78
thermal expansion coefficient, 98
thermodynamic pressure, 114
thermomechanical continua, 77
torsion, 20
total derivative, 53
traction, 29
transport theorem, 70
tridiagonal matrix, 157
turbulence, 113
turbulent cascade, 124
turbulent flow, 117

undeformed configuration, 51
unilateral compression, 97
upwind difference, 156

velocity circulation Γ_C, 121
velocity field, 51
velocity gradient, 59
velocity potential, 120
virial theorem, 68
viscoelastic material, 80
viscous stress tensor, 114
vorticity stretching, 123
vorticity tensor, 60

wave equation, 148
wedge-dislocation, 107

Young's modulus, 95